The Garden Explored

Mia Amato
with the Exploratorium

An Exploratorium Book

An Owl Book
Henry Holt and Company   New York

Henry Holt and Company, Inc.
Publishers since 1866
115 West 18th Street
New York, New York 10011

Henry Holt® is a registered trademark
of Henry Holt and Company, Inc.

Published in Canada by Fitzhenry & Whiteside Ltd.,
195 Allstate Parkway, Markham, Ontario L3R 4T8

Library of Congress Cataloging-in-Publication Data
Amato, Mia.
    The garden explored / Mia Amato and the Exploratorium.
—1st American Edition
        p.   cm.
    "An Owl book."
    Includes bibliographical references and index.
    ISBN 0-8050-4539-2 (pbk. : alk. paper)
    1. Gardening.   I. Exploratorium (Organization)   II. Title.
SB453.A636   1997   97-14328
635—dc21

Henry Holt Books are available for special promotions
and premiums. For details contact: Director, Special Markets.

First American edition 1997

Designed by Gary Crounse

Be careful! The experiments in this publication were designed
with safety and success in mind. But even the simplest activity
or most common materials can be harmful when mishandled
or misused.

Printed in the United States of America
All first editions are printed on acid-free paper. ∞

10 9 8 7 6 5 4 3 2 1

Contents

For Paul
—Mia Amato

Introduction

Welcome to the Accidental Scientist, a series of books created by the Exploratorium to help you discover the science that's part of things you do every day.

In *The Garden Explored* we investigate the science of soil and water, of seeds and seasons, of bugs and roots. We answer questions that will help you no matter where you garden. Why should you always water on a rising temperature? Why do you prune at a certain time of year? Why do plants root more readily if there's a willow branch rooting in the same water?

The Garden Explored is filled with experiments that you can try in your own garden. Create your own hybrids by cross-pollinating irises. Diagnose common garden ailments by examining a plant's leaves. Use your knowledge of a plant's inner plumbing to shape a tree with selective pruning. Get the most out of your growing season by mapping where shadows fall in your yard.

At the Exploratorium, San Francisco's museum of science, art, and human perception, we believe that learning how things work is not only fascinating and fun—but can also expand and enrich your experience of your favorite activities. Knowing that the texture of a plant's leaves indicate its water needs can help you choose the right shrubs for your yard. Realizing how plants absorb nutrients from the soil can help you fertilize more wisely. Understanding the wildlife that visits your garden can help you attract birds and butterflies and discourage snails and other pests.

Be warned: we've found that once you start noticing the science in gardening, you may find it difficult to stop. You'll find yourself noticing things that you never paid any attention to before and asking questions that you never thought to ask. Have fun!

Goéry Delacôte
Director
Exploratorium

Plants and People

1 Exploring the Natural World

I T IS A POPULAR fiction that people who write about gardens must have spectacular, gorgeous plantings in their backyards. Not me. My friends actually make jokes about my garden and call it "The Lab" because there are always so many experiments going on, some in beds, some in pots, some on the windowsill. This usually means that at least one area in the garden looks perfectly horrible, because a test I was running on disease resistance in old-fashioned roses has resulted in blobby, splotchy shrubs, or the peanut crop was thwarted by rummaging squirrels.

Nothing ventured, nothing gained—and the words, "You can't grow that here" are fighting words to me. True, I live in northern California, which supports a wide variety of landscape choices, but I've been told again and again my climate is too cold to grow blood oranges, and also that my climate is far too warm to produce flowers on the *Ume*, the Japanese flowering plum (*Prunus Mume*), which blooms in the snow in Tokyo. Well, my backyard *Ume* blooms just fine, thank you, and last year I did get three fruits from my 'Moro' blood orange, and they were very tasty indeed.

"Nobody sees a flower, really—it is so small— we haven't time, and to see takes time, like to have a friend takes time."

—Georgia O'Keeffe

If you are reading this book, I suspect you might also be the kind of gardener with an open mind and an itchy trowel. Perhaps you, too, have grown dissatisfied with the gardening advice that can be gleaned from books, radio call-in shows, and your own Aunt Fanny—"do this," "do that," "do it at this certain time." Just don't ask *why*, because "that's the way it is." Well, this book will tell you why certain things work and certain things don't, and will give you a scientific basis for backyard gardening that you can use with every plant you grow.

Gaining knowledge about the natural world allows you to adapt your gardening style, and manipulate soil, water, fertilizers, even animals and insects, to help your garden grow. You don't have to be a rocket scientist to understand the basic theories, and you don't need fancy equipment. Long before we had electron microscopes to dissect plants to the very core of their protoplasts, gardeners puzzled out many botanical theories using only their own five senses—what scientists today would call empirical observation—along with dirt-plain common sense.

Imagine, for instance, that you were a gardener living in Spain some 500 years ago. A friend drops off some seeds for you, from a plant he does not have a name for. Grow these seeds, says your friend, and you'll be surprised at what you get—a plant with delicate yellow flowers and the most delicious berries!

Sure you want to grow the seeds—but how, when you don't know what kind of plant it is? It helps when your friend tells you the seeds came from a distant tropical jungle, where the summers were baking hot and there was no winter snow. The climate was unbearably humid, and it rained a lot.

So you might try your luck with a scientific approach, waiting until your winter frosts were over to plant the seeds in the warmest, sunniest spot in your vegetable garden. You'd give the seedlings plenty of water. Perhaps you'd notice the plants grew lengthy shoots, but had no tendrils or suckers to cling with, so you might tie the vine up off the ground so the little fruits would not be damaged. And boy are those fruits tasty!

Tomatoes

And so you could successfully grow a tomato plant, even if you didn't know "tomato" was its name. Back in the 1500s, when the golden age of plant exploration began with Columbus's transit between hemispheres, gardeners were just as avid to try something new as we are today. They treasured every scrap of knowledge about new fruits and flowers, and learned by luck, trial, and error.

Modern gardeners might do well to imitate those adventurous gardeners of an earlier era, taking the added advantage of all we have learned since 1500. If you want to grow a better tomato in your own garden, it helps to realize the tomato's original home is the tropical jungles of Central America. Suddenly so much of tomato growing makes sense: why they wither in the cold; why surrounding the plants with heat-radiating black plastic mulch or a black rubber tire makes the plants grow faster; why honeybees often ignore their flowers, but bumblebees love them. (The bumblebees that usually pollinate tomatoes are, like the tomato plant, only native to the Western Hemisphere.)

Honeybee

Be More Than a Gardener—Be a Plant Explorer

The average suburban backyard contains what in 1500 would have been an astounding collection of plant species gathered from the four corners of the globe. Dahlias in the flower bed come from Mexico, while the yellow forsythia hedge hails from China, and the tulips along the walk are native to Turkey. Rhododendrons flanking the front door originated from the Himalayan snows of Tibet. That African violet on grandma's windowsill comes from African rain forests.

Knowing where a plant comes from is often the first step in a scientific approach to growing it, whether the plant is sweet corn or a giant sequoia. Give a plant what it wants, and what it is used to, and you can grow blueberries on the tenth floor balcony of a Manhattan high-rise, or orchids in your basement. (Books that pinpoint plant origins include Graham Stuart Thomas's *Perennial Garden Plants, or the Modern Florilegium* and other books in the same series, and Cornell University's *Hortus* plant dictionary, which, in its third edition, lists over 25,000 species and weighs seven pounds. (See Sources: A Resource Guide for Readers on page 146.)

Find the plant's point of origin and you've got a shortcut to cultural information on its sun, soil, and water needs. Our so-called "English" lavender (*Lavandula angustifolia*) is really native to the rocky cliffs of Greece; British gardeners know that to make it thrive they must give it chalky, rocky, poor soil. American gardeners who attempt to make an "English" flower border often blindly mimic the fertile and moist climate of Britain, and place their lavenders in rich, well-irrigated soils, and so they get poor flowers, if not root rot, for their efforts.

If you knew nothing about lavender except that it came from sunny Greece, you might have enough information to grow it

Lavender

Where in the world did that plant come from?

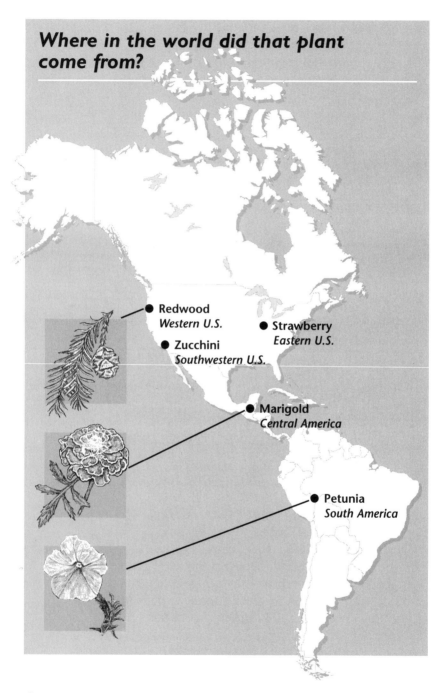

Redwood
Western U.S.

Strawberry
Eastern U.S.

Zucchini
Southwestern U.S.

Marigold
Central America

Petunia
South America

Plants that explorers brought back from exotic corners of the globe are now commonly found in suburban backyards. Understanding where a plant originated can help you figure out what that plant needs to thrive.

Apple
Asia Minor

Snapdragon
Mediterranean

Tea Rose
China

Wisteria
Japan

Lilac
Persia

Narcissus
Africa

Orange
Southeast Asia

Geranium
South Africa

Eucalyptus
Australia

quite well. Of course, reading a British gardening book, or any good gardening book, might give you the same information. But knowing where a plant is from may not be enough to help you grow it well in your particular spot on the globe. You must also remember where you are. As Lauren Springer, garden columnist for the *Denver Post* once observed, winter weather in the Colorado Rockies more resembles Afghanistan than England.

Growing plants from different parts of the world can be a difficult juggling act—but so much of the fun in gardening is meeting this challenge in creative ways. I once heard a great story about a Manhattan matron who bought a cactus plant to grow in a sunny window of her Park Avenue apartment. Then she ordered a subscription to a Tucson, Arizona newspaper. When the newspaper arrived each day she checked the Tucson weather report. And whenever it rained in Tucson, she watered her cactus.

Thanks to lawn sprinklers and watering cans, a climate factor such as rainfall is perhaps the easiest to manipulate at home (see Chapter 2). Many gardeners find themselves divided on whether to buy plants that suit their soil conditions (discussed in Chapter 3), or make the effort to adjust soils to suit plants they want to grow. With flowers, one of the most critical, and most misunderstood, factors is the number of hours of daylight. This is related to your latitude, where you are on the globe (more on this in Chapter 4). And if you want fruits, you have to arrange for pollinating agents, as America's early apple farmers did.

In 16th- and 17th-century America, apples were an important staple of colonist's early home farms. They were often the first trees planted, brought over on ships as rooted cuttings or small saplings. Besides the food value of the fruit, apples were made into hard cider, the most popular alcoholic drink in early America.

While the apple was foreign to the Western Hemisphere, related plants, such as mayhaw and serviceberry, grew wild. These wild plants had native insect pollinators that helped the apple trees set fruit until the colonists began importing European

Apple blossom

honeybees as well. Among the historical papers of our third president, Thomas Jefferson, are several references to bringing back bees and information from visits to hives and orchards during his tenure as ambassador to France in the 1700s. The role of bees in fruit production had, of course, been known to farmers for millennia: around 30 B.C., the Roman poet Virgil had a second bestseller (after the *Aeneid*) called *The Georgics,* which is a farming guide in verse that devotes a big section to the proper design and maintenance of beehives, even then deemed vital to field and orchard crops.

Jefferson's efforts to grow Virginia sweet corn in Paris are also well documented; his experiments succeeded since corn, like many grasses and grains, is not cross-pollinated by insects, but relies on the wind to move pollen from plant to plant. It rarely occurred to the early plant explorers to bring New World pollinators, such as the bumblebees native to North America, to Europe—at least until 19th-century naturalists such as Charles Darwin began to point out the amazing interdependent relations between plants and animals, a discipline we now call *ecology.* Many exotic flowers got reputations for being "too difficult to grow" simply because Europe lacked the insects to pollinate the flowers and create seeds in a garden setting. Some of those bad reputations are perpetuated to this day in modern gardening books.

By the Victorian era, once-rare tropical plants, such as the philodendron, were common fixtures in middle-class homes. Even

> From England I have this week received a new recipe for the plant that is sick or dispirited. It comes from Mollie Panter-Downes. A friend of hers in Surrey was showing an ailing wisteria vine to a gardening acquaintance. 'Oh,' said the visitor, 'all that wisteria wants is a nice rice pudding. They love them!' Accordingly, a rice pudding was made, cooked, well-sugared and laid round the feet of the vine, which promptly sat up, regained its tone, and is now full of health and pudding. There is probably a chemical explanation for this, but I would rather not know about it.
> —Katherine S. White, *Onward and Upward in the Garden*

the poorest families traded seeds and slips to grow on a windowsill or in a farm garden. But the era of new discoveries was far from over. In fact, it still goes on.

Ever hear of a fruit called shipova? Most folks in North America didn't know it existed until the end of the 20th century, when the dissolution of the Soviet Union released, like the water from a pent-up dam, cascades of exciting and different fruits and vegetables that Soviet agricultural scientists had been working on for years. Shipova (*Pyrus* x *Sorbus*) is a hybrid cross between a pear tree and a mountain ash that grows wild in what used to be Yugoslavia.

The collapse of cultural and political borders has been a boon to our flower borders, with waves of unusual bulbs arriving from South Africa (such as velthemia, a pretty purple spring companion to iris and tulips) and roses from mainland China that were unavailable for more than a century (though one that translates to 'Dragon with Pearl in Mouth' has been grown in Chinese gardens for 2,000 years).

Rose mallow

All this, of course, has meant a frenzy of plant buying, and an even more insane scramble for horticultural information excitedly swapped by hobby gardeners over the back fence and over the Internet. As in an earlier age, we continue to learn by luck, trial, and error. But since you are reading this book, you'll have one advantage over the punters—that certain edge you get when you understand basic plant science.

What's in a Name?

The discerning plant explorer can find out quite a bit of extra information by noting the scientific name of any vegetable or flower. Sometimes called the Latin name, it usually appears in catalog descriptions, along with the lovely common names we have. Forget-me-not, heartsease, angel's fishing rod, and love-in-a-

See for Yourself

1 Where on the earth do you garden? What latitude do you live in? Get a globe or a good world atlas and pinpoint the latitude lines nearest your city. Then follow the lines around the globe or across the atlas to find your "sister cities" in other parts of the world.

Odds are good that plants that grow in these foreign locations will be happy in your backyard. To find the best match, use other information in the atlas to find regions that have a similar elevation (mountains versus sea level, for example) and similar weather.

2 Plant a garden of weather-forecasting plants to help you gauge rain or dry weather as the ancients did. Include flowers such as scarlet pimpernel (*Analgalis arvensis*), a low-growing weed whose red blossoms only open in the morning if the day will stay sunny through the afternoon. This flower reacts to atmospheric pressure— the same thing a barometer measures. (A common name for pimpernel was "poor man's weather glass.") Morning glories also usually don't open on days that threaten rain.

Plants with trembling, waving leaves, such as ornamental grass or bamboo in a container, will alert you to high winds that may spring up on a summer afternoon, dry winds that suck moisture from the air and soil and leave your newly planted pepper plants bone dry. Rose mallow (*Hibiscus Moscheutos*) will close its summer flowers by noon on days when the temperature climbs to the point that lettuce starts bolting and pea pods dry to wisps. The mallow and grasses can remind you to water or re-water vegetable beds before your precious plantings get too parched.

Stick a few delicate but expendable flowers (cheap impatiens) in low-lying areas that will receive the first frost—then you'll have fair warning to cover up or bring indoors more valuable specimens. As a backdrop, consider aspen or poplar, if the site is open to wind. When the leaves of these trees show their silvery undersides, it's usually a sign that a storm is on the way.

puff (so called because the black seed is imprinted with a white heart, and sits within a fluffy seed case) are some of my favorite common names.

In the mid-18th century, Swedish botanist Carl Linnaeus devised the form of plant classification we use today, which is known as the *Latin binomial system*. Latin was (and still is) the language of scientists, so Linnaeus used Latin words—not Swedish words—to classify the plant families we use today, with more detailed classifications underneath them, termed *genus*, and even more detailed divisions under them, called *species*. Naturally occurring variations within a species, such as double flowers (*flore-pleno* in Latin), were identified with extra descriptive words when needed, along with the names of selected varieties, what botanists call *cultivars*.

> *"I do not believe in pampering plants. If they are miffy, let them go."*
> —*Elizabeth Lawrence,* In a Secret Garden

Latin words in plant names usually highlight that plant's best feature or most particular trait. Some are easy to guess: *pendula* means pendulous or hanging; *glauca* means gray-colored. So a tree described as *Cedrus Atlantica glauca pendula* describes the weeping blue Atlas cedar. The genus, *Cedrus*, identifies the plant as a type of cedar, normally a tall tree; the species, *Atlantica*, means it is one of the cedars commonly found on the coast of the Atlantic Ocean; *glauca* means the leaves are grayish (the color is actually quite a pretty powder-blue); and *pendula* means the branches hang down in a weeping form.

That's lots of information for a gardener who might be looking for a handsome landscape tree to put in the front yard. Sadly, though, many eager gardeners simply see a pretty blue tree in a nursery container and plant it at home without even looking at its label and considering all these good clues. So the weeping blue Atlas cedar winds up jammed against a house foundation, pruned to ugliness in later years as the homeowner attempts to keep its naturally tall inclination from squishing against the roof over-hang, its needles sickly and pale from the lack of airflow, when it really wants to be in a stiff wind, if not an ocean breeze.

Cracking the Code

Finding your way through the jungles of botanical Latin can teach you a great deal about the plants themselves. Take "baby's breath," a poetic name that beautifully describes the plant's airy cloud of tiny, white blossoms. By also learning its Latin name, *Gypsophilia paniculata,* you will better remember that the plant grows in calcareous soil (*Gypsophilia* means "gypsum lover") and the flowers are arranged in *panicles* (loose, spreading flower clusters). Another species, *Gypsophila repens,* is more low growing, since *repens* means "creeping." In the end, learning botanical names will make you more than just a better shopper; it will make you a better gardener and increase your understanding of the natural world.

—Barbara Damrosh, *Gardener's Latin: A Lexicon*

Some Latin Terms for Landscapers

aromaticus: fragrant

cordifolius: with heart-shaped leaves

elatus: tall-growing

fastigiatus: upright or column-shaped

frutescens: shrubby, low-growing

millefolia: millions of tiny or finely divided leaves

montanus: native to mountain sites

paniculatus: flowers in loose clusters, or panicles

praecox: early-flowering, as in precocious

nutans: nodding or arching shape

pendulus, pendula: weeping shape

repens or *reptans*: creeping, as in reptile

spicatus: spiky flowers

tenuifolius: thin leaves

tragophyllus: leaves with a goat-like odor

—Bill Neal, *Gardener's Latin: A Lexicon*

Read the labels; the clues are there. Sometimes the second half of the Latin name actually tells you where the plant is native. Anything labeled with the species name *canadensis* would be from North America, for example, and is probably hardy enough to grow in Canada. So hey, you can probably grow it well in Maine or Vermont, too.

Sometimes a species name is a person's name, the person who "discovered" the plant, or to be more precise, found the species wild and brought it back to be propagated through the nursery trade. Anything tagged *wilsonii*, for example, is likely to be native to the interior of mainland China, because that was the stomping ground of Ernest "Chinese" Wilson, who popularized such garden goodies as forsythia.

The plant-hunter's custom of naming a genus after friends explains some of the non-Latin genus names, such as *Fuchsia* (for a Mr. Fuchs). When Lewis and Clark prowled the Rocky Mountains in search of the Northwest Passage, the botanist they brought along named two of the prettiest wildflowers he found, a peppermint-toned alpine (*Lewisia*) and a brilliant-hued annual (*Clarkia*), after the leaders of the expedition.

Follow the Families

If you want to be a real plant explorer, you won't stop with reading the Latin names on plant tags. A great way to expand your gardening horizons is to move up a step higher and learn to recognize plant families.

In practical terms, Linnaeus's codification of plant families is a great shortcut in finding new flowers to grow. For example, a lot of plants with daisy-type flowers grow in similar conditions, prefer-

Shasta daisy

ring the airiness of open fields and plenty of water to their roots. It turns out that all daisy-looking flowers are more or less genetically related in the same plant family, the *Compositae*.

If you can grow one kind of daisy well, you can usually grow another with similar techniques. Any garden site that suits an aster

(*Aster novi-belgii,* native to North America) will usually suit the ground-hugging English daisy (*Bellis perennis*) if not the South African gerbera (*Gerbera Jamesonii*), which is frost-tender.

If you look closely at apple and strawberry blossoms and then at wild roses, it is easy to see the family (*Rosaceae*) resemblance in the five-petaled flowers with their centers of golden stamens. Just as members of human families may share the same allergies, all rose-family members can suffer a fungus disease called "rust." While some rusts are plant-specific, rose-family members in your garden can spread this disease to each other. A gardener who religiously sprays apple trees in their winter dor-mant season to control rust must remember to spray the backyard rosebushes, too.

Broccoli

With vegetables, a general knowledge of family resemblance is important, since plant families typically share not only soil and sun pref-erences but also the same insect pollinators and insect pests. Carrots and the more exotic salsify will both attract the carrot maggot fly, as will Queen-Anne's-lace, because they are all in the carrot family (*Umbelliferae*). Carrot-family members also attract swallowtail butterflies, so gardeners can plant carrots and their relatives to make a but-terfly garden. (See Chapter 6 for details.)

Other important vegetable families are the onion group (*Amaryllidacae*), which includes garlic and daffodil—two popular garden dwellers that, like all onions, prefer wet springs and dry summer soils, and the mustard family (*Cruciferae*, alternately called *Brassicaceae*), which links broccoli, cabbage, and that trendy red salad ingredient, radicchio. In the garden, all these mustardy plants must be protected from the greedy green larvae of the cabbage moth, a white butterfly that can be controlled with a nat-ural bacteria. (More on that, too, in Chapter 6.)

Today's gardeners often sadly ignore these useful family ties. But in older times, even the most unlettered peasant could recog-nize a member of the mustard family, because the flowers of all mustard-family members have exactly four petals, in the shape of a cross. Popular tradition among Old World Christians held that this was a sign from God that all such plants were healthy to eat, and that a plant with four flower petals in the shape of a cross is never poisonous. Botanists will tell you that there are indeed no poisonous members of the mustard family; all plants classified as *Cruciferae* have a similar chemical/genetic structure that includes no mammal poisons.

But if people still gardened only by tradition, none of us today would be enjoying the juicy and fantastic taste of sun-ripened, home-grown tomatoes—since you can't always use a family resemblance to determine if a plant is edible. If you had been that gardener in circa–1500 Spain, and your neighbors had seen you

Slaves to Fashion

The Dahlia's history is strange. Its origin can be traced to Mexico, where the Aztecs called it *Cocoxochitl,* and, in the year of the French Revolution, tubers were sent over to a French priest who was chief gardener at the Escorial....

From seeds or tubers—it is not known which—Napoleon's Josephine managed to grow Dahlias at Malmaison, and she was exceedingly jealous of her collection. Only her hands planted and weeded and generally cared for the imperial Dahlias, until she had so many plants that it was necessary to put a gardener in charge of them. Inevitably, corruption followed. One of her ladies-in-waiting who had asked for a tuber and received a curt refusal was determined to outdo her royal mistress's collection. Her lover, a melancholy Polish prince, was ordered to steal sufficient tubers from Malmaison. He did not attempt it himself, but instead bribed the gardener to carry off a hundred roots.

When she heard of it the Empress was outraged. She dismissed the gardener and the lady-in-waiting, exiled the melancholy Pole from court, had all her Dahlias chopped up and dug in, and never again would hear the plant named in her presence.

—Tyler Whittle, *The Plant Hunters*

eating the fruits from the tomato seeds you had sown, they prob-ably would have jumped over your garden fence yelling and screaming, and calling for a doctor and a priest.

By looking at the fruit, vines, and leaves of the tomato plant, your neighbors might easily have guessed it was in the same plant family as the poisonous deadly nightshade. And they would be right, since both tomato and deadly nightshade are in the same plant family, *Solanaceae.* In fact, well into the 18th century, Europeans still regarded tomatoes with great suspicion and considered the fruits poisonous—though they were grown in many flower gardens as an exotic ornamental.

Tomato

Perhaps it is to Linnaeus that we owe the joy of eating toma-toes. Under his scheme, deadly nightshade and tomato still share the same plant family, *Solanaceae,* but are differentiated by their Latin binomials as *Solanum Dulcamara* for deadly nightshade, and *Lycopersicon esculentum* for the edible tomato. The Latin word *esculentum* means "edible" or "tasty." *Dulcamara* means "bitter." But knowing the family provides a further hint to the cook as well as the gardener, because the leaf of a tomato plant is just as poisonous as the leaf of any nightshade.

New and Improved

An understanding of Latin-based plant classifications is the first step to understanding the science behind *hybridization*—the art of creating new plants from old. Whole libraries full of science books have been devoted to systems for hybridizing plants, but all a gardener really needs to know is how to tell the players apart.

Latin binomials make it all so simple. A *hybrid* is the offspring of two plants that are genetically compatible—in some way related through the linkage of species, genus, and plant families.

It's extremely easy, for example, to cross between two plants that belong to the same species. If you have a blue bearded iris

Hybridizing at Home

I f you are a patient gardener, consider growing your own hybrid. Bearded iris, a very common garden plant, is easy to work with, since the plant parts (male stamens, female pistils) are big and easy to find. The flowers are also quite large, so color variations in the blossoms can be noticeable, and a plant will grow to flowering size in 3–5 years.

The first step is to pick your female parent plant, and cut off the male parts (stamens) on one of its flowers before pollen has begun to appear. Then you'll need a male parent plant with lots of pollen. You can cut off the entire flower from the male parent and simply dust it against the pistils of your chosen mother—or you can transfer the pollen with a soft paintbrush.

Bearded iris

Hybridists will traditionally cover the mother plant with a paper bag to prevent any other pollen from bastardizing the flower. Bagging the bloom also makes it easier to collect the seeds when the flower has matured. A most important step, often neglected by amateurs, is to label the stem of the pollinated flower with the names of both parents. Tie-on tin or copper labels will hold up the best, and you can use the same label to mark the flat where you grow the resulting seeds.

pistil

stamen

Other garden flowers easy to hybridize are daylilies, true lilies, and roses. Rayford Reddell's rose book (see Sources) has a very easy-to-follow, step-by-step guide to producing rose hybrids at home.

Will fame and fortune follow? Just keep in mind that professional rose growers often do 100,000 crosses before they wind up with a hybrid worth offering for sale.

and a white one (both *Iris x germanica*) and cross-pollinate them, the resulting seeds may be plants that bear flowers of an intermediate hue. Cross-pollinating iris is so simple you can do it at home: you simply act like a bee, and transfer dusty pollen from the male part (*stamens*) of one iris flower to the female part (*pistil*) of another iris flower.

Crossing between two different species within a genus is a bit harder. Sometimes the incompatibility is physical: the pollen may arrive at the pistil, but the pollen grains may be too big in size to travel down the pistil's little tube to the flower ovary, where the waiting egg will make the seed. A little more manual labor is necessary when working with the parent flowers, but that's how Luther Burbank invented the plumcot, a plum-apricot cross, a tree that fruits as easily as a plum, but tastes more like the harder-to-grow apricot. Both share a genus: Burbank worked with the Japanese plum (*Prunus salicina*) and the apricot (*Prunus Armeniaca*). Botanist Floyd Zaiger has recently continued this work to produce the pluot (¾ plum, ¼ apricot) and the aprium (vice versa).

"Without careful selection each year all seeds will degenerate, so universal a rule it is that all things naturally tend to become worse."

—Virgil, The Georgics

Burbank also crossed among genera in the same plant family. A popular result of his work in *Compositae* is the big white Shasta daisy (*Leucanthum x superbum*), which was the result of uncounted crosses between Japanese chrysanthemums and a wildflower from the slopes of California's Mt. Shasta.

Sometimes a hybrid seed produces a plant that is sterile—witness the seedless watermelon. Sterile plant hybrids are sometimes called "mules"—a reference to the practice of breeding two animals of different genera, a horse and a donkey, to create a sterile animal that cannot reproduce, but is otherwise useful to its human masters. In the garden, mule marigolds are common. They're the ones that seem to keep on blooming and blooming forever, because the poor things just can't set seed.

Plant Explorers Wanted

Y ou don't have to be Luther Burbank to come up with a brand-new plant. There is plenty of room for new plant explorers. There is even an organization of amateurs, North American Fruit Explorers (NAFEX), who prowl wildlands of our continent in search of wild fruit plants, such as pawpaw, maypop, mayhaw, edible cactus, and wild plum.

NAFEXers don't expect to find a new genus, or even a new species. But sometimes they do discover, in wild areas, a pretty good specimen that has bigger or tastier fruit. A plant with promise can then perhaps be propagated by cuttings or divisions, and given to other gardeners—it is now a "selection" and can be given a cultivar name. If enough other people like it, it can enter the nursery trade.

I've met several people who have found and propagated ornamental selections of wild plants just in the past few years. One of them is Roger Raiche, the native plant specialist at the University of California's Botanical Garden in Berkeley. An avid gardener as well as a native plant fan, he brought home a bit of wild grape (*Vitis californicus*) whose leaves he discovered turned a bright and brilliant scarlet each fall.

California is, of course, not New England, and anything with brilliant autumn color attracts attention. After many requests from friends, he began propagating it. Now he has turned some over to commercial nurseries, and you can buy it mail order, listed in catalogs as *Vitis californicus* 'Roger's Red.'

Garden catalogs rarely use the term mule, but frequently tout the fact that a certain new plant is a hybrid—a way of trumpeting that this petunia or cabbage is new and improved. F1 hybrids are the result of a single cross; those listed as F2 are second-generation hybrids bred from non-sterile hybrid parents.

You may come across hybrids identified as *diploid* or *tetraploid* hybrids. These are mutant plants that have double, triple, or quadruple sets of chromosomes on the DNA strands of a single cell. (Plants, like people, usually have just one set, or pair, of chromosomes on a DNA strand.) The method used to make these *polyploidy hybrids* is to dose the parent plants with *colchicine,* a chemical derived from a species of crocus. When botanists crossbreed among the mutants, the final result can be seed-offspring with very different and often spectacular flowers. New varieties of garden lilies are often ploid-type crosses.

Sometimes scientists try to insert new genetic material into a plant's DNA strands using a "gene gun," a laboratory device that shoots a tiny tungsten bullet (I'm not kidding) coated with bits of plant cell tissue into a target of some other plant's tissue (usually a leaf). With luck, the tissue shoots right smack into the nucleus of one of the cells in the target leaf, and the jolt of impact causes the bits to bond, thereby creating a mutant cell. Other cross-breeding techniques include radiation treatments to induce mutation, or hormones such as *gibberellic acid* (see Chapter 5) that alter genetic structures.

> *"Roses are bisexual. The selection of who squires and who receives is arbitrary; the female is the rose that remains on the bush, and the male is the one tossed after use....Prepare the female first, in case you break it off the bush, in which case it can serve as the male."*
>
> —*Rayford Reddell,*
> Growing Good Roses

The downside of hybrids—especially those produced by laboratory methods—is that seed-offspring of hybrid crosses rarely repeat desired traits. This is why in some old yards a stand of once–vibrantly colored phlox seems to "revert" to sprouting only flowers in muddy purple. Purple is a dominant gene in phlox, and what's happened is that off-colored offspring have crowded out the original hybrids, which may have died out over time.

Parent plants of vegetable F-1 crosses are usually special selections, proprietary and not available as seeds to the general public. In other words, you can't make the same F-1 hybrid at home. You have to buy new seeds each year, and the seed companies make a little money off you each time.

There's been a lot of fuss lately in gardening circles about open-pollinated flower and vegetable seeds, which are produced by natural garden conditions—or in the case of commercial seed houses, by growing vast quantities of specific flowers or vegetables in large open fields. Genetically stable, a packet of such seeds will produce plants that, in turn, will produce a second generation of seeds almost exactly like the first. (Slight variations noticed in growing fields are "rogued out," or removed by field workers to assure uniformity.)

> "No occupation is as delightful to me as the culture of the earth, and no culture comparable to that of the garden. Such a variety of subjects, some one always coming to perfection, the failure of one thing repaired by the success of another."
>
> —Thomas Jefferson

In the centuries before hybrids, gardeners and farmers would let their favorite vegetables and flowers mature in place, trusting natural pollinators to help set the seeds they would save for replanting year to year. Open-pollinated seeds have become popular again for two reasons: many find it cheaper to save their own garden seeds from year to year, and some gardeners want to grow historic or "heirloom" open-pollinated flowers or vegetable varieties. (See Sources.)

Hybridists are also extremely interested in these older, passed-down varieties as genetic raw material. Science does march on, and you should remember that many open-pollinated vegetables passed out of favor because newer hybrids were less hassle to grow. Hybrids are created not just for bigger plants and nicer flowers, but to breed in disease resistance, a big plus if you are trying to grow tomatoes in ground that may harbor soil-borne blights, such as verticillum and fusarium wilts. The celebrated Quaker heirloom tomato 'Brandywine,' for example, has a great reputation for flavor, but also a reputation for becoming a puny, stunted, catfaced, blossom-rotted, blighted old stick in some people's gardens.

Full Circle

Fads and fashions in gardening come and go as quickly as the seed pods on a portulaca. The urge to show off for the neighbors by planting something new and different is just as strong in us as it was to the greedy Empress Josephine, who, while Napoleon was battling in Europe, dispatched her own spies behind enemy lines—not to bring back military secrets, but just to acquire new varieties of cross-pollinated rosebushes to grow in her garden at Malmaison.

Today we have the fad for "native plant gardens," with a rush to hybridize promising North American fruits roundly ignored for 500 years, such as the pawpaw and serviceberry. We can have "heirloom gardens" that include open-pollinated vegetable seeds and the very same roses smuggled into Malmaison—intriguing old varieties we can enjoy anew for their fragrance, color, and form.

The first recorded instance of a plant actually being dug up and transported for somebody's garden is 1459 B.C., when Queen Hatshepsut of Egypt (who sometimes sported a false beard in her court appearances) sent a collecting expedition to the land of Punt (now Somalia) to bring back trees whose sap could be used to make incense. Hieroglyphics that can still be read nearly 3,500 years later record that 31 trees survived the journey, their roots packed in baskets of soil carted down the Nile by boat, and were planted at Thebes near the Temple of Amun.

So grow what you want; give everything you can dream of a try. Give in to the temptation to plant something different, difficult, new, or rare.

It's the human thing to do.

Inside Plumbing

2 Understanding the Vascular Nature of Plants

THEY SAY HUMAN BEINGS
are 98 percent water. I've been
told a cabbage leaf is 92 percent water, and I know you can
sometimes actually see water coursing through the rib of a leaf of lettuce.
But even when you can't see water flow-
ing in a living plant, water is always there. There's
water in the cells of a Norway maple when it's
dormant and bare-leafed, seemingly only a dry stick
under winter snowfall; a desert cactus resting
through a torrid Arizona summer has water in
all its thick-walled cells as well.

Because water has, at its molecular level,
an ability to dissolve minerals like iron and calcium, and also the ability to carry along gases
such as oxygen and carbon dioxide, plants rely
on internal plumbing—and the natural movement of water—to run their growth functions.
Roots, porous as a sponge, absorb water. The water

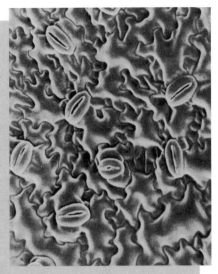

his micrograph shows the microscopic air vents—known as stomata—on the surface of a leaf.

On the common jade plant, you can see the stomata if you have a magnifying glass, some acetone (nail-polish remover), and a brittle piece of plastic. (The clear plastic lid from a yogurt container is perfect.)

Peel off the back of a large leaf, removing the leathery skin to reveal the pulp underneath. Soak the leaf back in acetone, shake off the excess, and lay the leaf pulpy side down onto the brittle plastic. Leave it for one minute, then remove the leaf. The acetone will etch the plastic, showing a pattern of stomata from the leaf. The geometry is quite beautiful when viewed with a magnifying glass.

travels up the stems; excess water is evaporated back into the air through microscopic air vents (called *stomata*) on the surfaces of their leaves, in a process that is called *transpiration*.

The movement of water in plants is initiated not by the roots but at the leaf level. Once transpiration begins, a simple suction pulls water upwards all through the day, from the roots to the tips of each leaf. If you forget to irrigate a plant during a warm, dry period, water evaporating through the stomata causes the roots and stems to suck up all available soil moisture, until there is nothing left for the plant to suck up. Without water, the plumbing structure collapses like a puckered drinking straw, and the plant wilts.

Leaf transpiration is like sweating: on a bright sunny day, radiant energy from the sun's rays stimulates leaf stomata to open, and this makes a plant transpire—not unlike the way heat from the sun makes you sweat. Guard cells around a plant's stomata expand

and contract through the day, responding to available sunlight. In most plants, transpiration begins as the sun is rising and increases as the sun shines high in the sky.

If you imagine that a leaf acts like your armpit, a stem acts like a drinking straw, and that roots act like a sponge, you will have a pretty good idea of how complex and busy the vascular activity can be in a petunia that just seems to be sitting there.

Once you understand the internal plumbing of plants, much of good gardening suddenly makes sense. If you water a plant in the early morning when the stomata are opening, the resulting transpiration will draw water up into the plant and then the plant's own internal structure can do an efficient job of making sure the liquid refreshment pumps through to all parts.

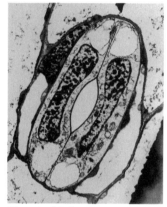

But if you water in the evening, as the darkening sky triggers stomata to close, the upward movement of water slows and liquids stay down around the

Stomata detail

root level or drain uselessly into the ground. That's inefficient. Splashing water on leaves and flowers as night draws on is even worse. Wet leaves and stems don't have a chance to dry at night, and they become an open invitation to fungus diseases, such as downy mildew and rust.

Y ou can judge a garden by the state of its phloxes. They are a pitiable sight when parched, a living reproach to their owners, to the country's water resources and, in the background, to an entire social structure which fails to provide an environment fit for phloxes, and to an inadequate educational system which fails to din into obstinate blockheads that an adequate water supply is essential to good gardening.

—Christopher Lloyd, *The Adventurous Gardener*

When to Water

Test your awareness of when to water. Are these three statements true or false?

• Always water on a rising temperature.

True. Leaves open their stomata in response to increased sunlight, drawing in radiant energy that we humans feel as increased air temperatures. If you irrigate a garden or a lawn about an hour before sunrise, you'll be catching the transpiration period at its most active, and efficiently moving water from your hose or sprinkler into the roots and stems of your plants. An added bonus: when you water in the morning, leaves and flowers have a chance to dry off slowly, which helps prevent diseases.

• Never water at noon or in the full heat of the day.

Also true. Irrigating lawns with a sprinkler at midday wastes water that will simply evaporate in the air before it hits the ground. Sprinkling plants in the hot sun sometimes makes water droplets act as little magnifying glasses, and plants may be burned by sunlight before the droplet is absorbed or dissolved. This "sunburn" usually shows up as white or brown spots on leaves. (African violets in a sunny window are especially susceptible.)

• It's best to water your lawn at night.

False. Plants close their leaf stomata during periods of darkness. In the daytime, flat blades of grass are champions at transpiration —that's one reason why a grassy area always feels cooler than pavement on a hot summer day. But don't run sprinklers at night, when the water will just puddle around roots or drain away unused. Water at night and next thing you know you're getting little brown mushrooms and slimy green moss between the grass blades; that powdery orange color in patches on your lawn will be rust (a fungus disease that disappears with drier conditions).

Revive a Wilted Tomato

Garden plants accidentally left to wilt on a very hot day will usually recover if given quick attention. First, "prime the pump" by sprinkling the leaves of the plant with water. The leaves will absorb a bit of moisture through their now gasping stomata.

Then soak the ground, slowly and thoroughly, using a fine spray or a slow trickle. Use your finger or a stick to make sure the soil is being soaked to a depth of six inches at least, to revive roots near the soil surface.

Tomatoes and most plants will respond to this treatment. In response to drought stress, flowers or fruits may drop off; leaves that stay yellowish or crispy-brown can be trimmed off with scissors once you see new growth.

The Right Amount of Water

You can use your knowledge of plant plumbing to know when to water, but how do you know how much to water?

How much water a plant needs depends on how much it transpires or evaporates water, and that, in turn, depends on your daily local weather. A plant transpires more rapidly when the weather is sunny than when it's cloudy—though plants transpire even during rainy weather. A plant's water requirements also depend on its location in your yard—a petunia that's baking in bright sunshine needs more water than one that's under a tree's dappled shade.

Agricultural scientists concerned with frequent droughts in the western U.S. have been trying for several years to determine exactly how much water is needed by, for example, a growing rice plant or a stem of broccoli. They've found that a single cornstalk can use up to a gallon per day—less when it's young and small, more when it's taller and has more leaf surface. A large apple tree with ripening fruit, on the other hand, needs fifty gallons of water per day.

Calculating in the water lost from air and soil surface evaporation with the amount lost via transpiration and plant growth, agricultural scientists have come up with figures called *evapotranspiration rates* (ET for short) for some crops. Some farmers use ET rates to determine how much water is needed to irrigate their fields.

Robert Kourik, author of several books about sustainable agriculture for small farmers and home gardeners, has suggested the goal of the home gardener should simply be to replace exactly the amount of moisture that is lost each day, then add just a little more. (He calls this "topping the tank.") To manage this, he often suggests drip irrigation systems that release a daily dose of water, sometimes by the teaspoonful—exactly the kind of shallow, frequent watering most other garden book writers sternly counsel against.

His technique works well for growing vegetables, which have shallow roots, and for flowers grown in pots. To apply ET rates to larger landscape plants, I recommend his excellent book along with the explanatory pamphlets written and published by the University of California. (See Sources.)

Of course, you can't garden strictly by the numbers. The very best way to figure out how much water you need to provide is to study your plants. A dry plant will have dull leaves or limp ones; a happy plant has leaves that are stiffer and bright. A dry pot is easier to lift than a soaked one, and when you stick a pencil into the ground around your grass or vegetable beds, it should come up damp at least halfway.

My favorite shortcut for figuring out if I'm watering enough is to include strategically placed indicator plants in my vegetable and flower beds. All plants will wilt if they're not getting enough water, but some wilt faster than others. Pineapple sage (*Salvia elegans*), with its aromatic foliage and bright red flowers, is a first-class drooper when the ground starts getting dry; Mexican bush sage (*Salvia leucantha*) folds its leaves and sulks when deprived of water. Among shrubs, hydrangea is one of the first to wilt; it's a good choice for this monitor role in patio containers. Both lettuce and leeks are vegetable bellwethers to warn when too dry is too much.

ET and CIMIS

alifornia farmers can check in to California Irrigation Management Information System (CIMIS) to find out exactly how many inches of water to put on their crop on any given day. Home gardeners can also get this information from their local cooperative extension office or by calling the CIMIS hotline at (800) 922-4647.

If you'd like to use this information, you'll need to estimate just how many inches of water your sprinkler supplies in a given time. To determine this, just put a coffee can underneath the spray for a given period (half an hour, for example). Then measure the height of the water in the can with a ruler. If you have a rain gauge, you could use that, but coffee cans are considered the universal rain-gauge substitute among agricultural scientists.

If you are using a drip irrigation system, your emitters all have flow rates given in gallons per hour. On an average 12-inch by 18-inch spacing, an emitted gallon delivers approximately one inch of water; the actual depth of water-soil penetration depends on how porous the soil is.

According to Richard Snyder, biometeorologist at the University of California at Davis, evapotranspiration (ET) rates don't differ greatly from crop to crop as much as they do from place to place—local weather and time of year being the most significant factors. To set up CIMIS, baseline ET rates were established for pasture grass at locations throughout the state.

Farmers adjust the grass figure up or down for certain crops: newly planted lettuce, with little leaf area, will need less water; a tall planting of mature corn needs a bit more.

"If you're growing a tomato plant in California's Central Valley, it would need about a quarter-inch per day," says Snyder. That's the average ET rate for the region in the summertime. Home gardeners, he notes, could apply this amount daily, perhaps with a buried drip irrigation system. Or one or two good waterings per week might suffice: "About two inches per week would be more or less right."

But if you would like to get some data about a particular plant's water requirements before you plant it in your garden, you can find some good clues simply by looking at its leaves.

To seal in some existing water within the leaf, and to prevent too much water from being lost, all leaves are covered with a protective coating called the *cuticle*. In deciduous species such as maples, the cuticle can be quite thin, and the leaves may feel soft and delicate. Maples evolved as forest plants, so they're used to being protected from harsh sun by other, larger trees. A single maple set into a bright sunny lawn all by itself will often get sunburnt leaves—crispy brown tips or white splotches—if it loses more water each day than it is taking up.

Species that evolved in harsher sun conditions—sometimes gardeners refer to them as "drought-resistant" plants—have evolved extra protection. Gray-leafed plants from regions with Mediterranean-type climates, places that receive very little summer water (California, South Africa, Greece, Australia) may have fuzzy, furry cuticles to shield the stomata and cut down water loss. Evergreens, such as redwood and pine, adapted to the bright air and windy slopes of higher elevations, have evolved thin needles, which transpire less than wide leaves. The needles have cuticles that also seem waxy to the touch.

If you are a gardener, this is all very important when you are designing and adding new plants to your yard. Gray-leafed Mediterranean species, such as lavender, bearded iris, or Maltese cross (*Lychnis coronaria*), will thrive in a bright, hot location, while woodland species, such as violets and columbines, will never be happy unless you also plan to set a brand-new shade tree above them.

To choose the right plants at your local garden center, all you have to do is feel for the leaf texture: if it's soft and tender, it probably prefers some shade; if the leaf is stiff or furry, it will grow well in harsh sun.

Because leaves and leaf-tips are the tag end of the water chain, any disruption or blockage of vascular flow shows up there first— at the end of the line. In the same way that too-dry conditions are first seen as a limp and wilted leaf, astute gardeners can learn to "read" leaves to diagnose plant diseases that may be attacking the stems and roots.

One of the most common mistakes gardeners make is over-watering their garden plants. Air in the soil is critical, and flooding the soil or a plant in a pot replaces the oxygen between soil particles with water, which can suffocate the plant along with beneficial animals (such as earthworms and helpful soil bacteria). With a houseplant, one of the first signs of overwatering is when the tips of the leaves start to turn brown. That's usually when the gardener says, "Oh no, the leaves are brown, the plant probably needs water!" and then proceeds to compound the problem by pouring even more water into the pot.

Savvy gardeners should also be aware of the differing water requirements of specific plants. A cholla cactus living in Santa Fe needs only an inch of rainwater per year to survive, though to bloom its pretty, papery magenta flowers it would prefer a bit more. A well-rooted Valencia orange tree planted next to the cactus needs about 25 inches of water per year, though it would probably fruit better if it lived in Ft. Lauderdale, where frequent tropical rains and humid air would suit it, too.

Drought-tolerant plants, such as juniper and wild ceanothus, perish pretty quickly from root rots if they're watered in the same style as thirsty roses and grass lawns. That's why flowers said to need "sharp drainage" or "well-drained soil" should be segregated from their thirsty brethren.

What's Going On Down There?

All this advice about watering is directed to one end—getting the right amount of water to the roots at the right time and in the right place. Unlike animals, a plant can't move to find a stream or river to drink from. Instead, it sends out roots to find water. Lots of them. I'm told a scientist once laid out and measured all the feeder roots found on one healthy tomato plant—and the total combined length came to about two miles!

When a seed germinates, the first thing that happens is that the plant sends out a root to look for water. This initial root usually becomes the taproot, which anchors the plant and prevents it from

When vegetables and fruits reach maturity, it's time to wean them slowly off water. Tomatoes, peaches, melons, strawberries, and grapes become sweeter as a lack of water concentrates sugars in their ripening fruits. Plants harvested for their essential oils, such as lavender and other herbs, should have their irrigation tapered off as the flowers appear; the scent will be more concentrated and will last longer.

being torn out of the ground by wind or varmints. Once the taproot sets a general direction for the root system's future underground activities, it begins to sprout secondary roots and feeder roots that may be finer than a hair, and these take over the work of finding and absorbing water.

Forget what you've heard about deep-rooted trees—most water-gathering activity takes place in the top two feet of soil, using the tiny feeder roots. Where conditions are perfect, the root range of an oak or similar large tree will extend to an area that reaches only about as far as the shade canopy of the tree. Landscapers call this demarcation the *drip line,* and it's the area where root growth is more active. When professional gardeners want to fertilize big landscape trees, they fertilize around the drip line instead of close to the trunk.

If conditions are less than perfect, a tree may send its feeder roots far afield to find water. Trees that are especially water-greedy, such as willows, will head their roots straight into a sewer line or septic tank; never mind that a house or street may be in the way.

Where a tree does not need to be stabilized (for example, when it's in a squat container on your deck), the taproot can be removed entirely without harming the tree. However, if a tree's feeder roots are disturbed, which can happen if you dig or pave too close to a tree trunk, the tree may die.

Some species of plants seem to grow bigger roots than you might think they need, looking at the size of the plant that's above the surface. Tomato plants, for example, prefer a root run that is at least 2 feet deep—any shallower and their growth becomes stunted. This only seems like a paradox because we consider the tomato plant a short-lived addition to our kitchen gardens. In its natural state (as we learned from Chapter 1) the tomato is a long, sprawling vine and a perennial at that. Is it any wonder it demands the same root space as clematis or a climbing rose?

Lawn grass is happy with six to eight inches of good soil for its roots to roam, though in sandy, sparse soils it will send roots down two or three feet to survive. On the other hand, the Phoenix palm is a shallow creature: it may grow to 40 feet in height, but rarely needs more than a plot measuring 5 feet by 5 feet by 5 feet to thrive.

Understanding what's going on beneath the surface is particularly important when you are transplanting. When a plant's roots are stressed, it can temporarily lose the ability to draw up enough water to replace what it loses through its leaves during transpiration. This is what causes sudden wilting when a tree or shrub is transplanted during hot, dry weather.

"When teaching a drip irrigation seminar, I stress maintaining good soil moisture in the top 2 feet, and I tell people not to worry about soaking any deeper.... Watering should be a function of how roots really grow, not how we imagine their growth."

—Robert Kourik,
Drip Irrigation for Every Landscape and All Climates

Deeply soaking the ground after transplanting helps, but it's always better to wait until a tree is leafless or until the weather can be counted upon to be cool and cloudy for a few days before you transplant a large tree or shrub. That's why the traditional period

T hese newly transplanted Phoenix palms can survive life in a concrete oasis because their roots are shallow. Subterranean irrigation lines support these street trees in the manner of a naturally occurring underground desert spring. The palms' fronds are temporarily tied into topknots to cut down the expanse of leaf surface open to dry air and sunlight, which would cause the plants to transpire too rapidly, drying out the plant before the roots have recovered from transplant shock.

for transplanting—everything from roses to fruit trees—is late autumn or early spring. Deciduous species are leafless then, and weather patterns reliably include cloudiness and rain.

Landscape professionals do transplant even big trees all year-round, and part of their aftercare is making sure a plant is watered deeply and frequently until its roots have recovered. They've also got a little secret that comes in a spray can: anti-transpirant sprays, such as Wilt-Pruf, derived from colloids found in seaweed. Sprayed on a plant's leaves, these products temporarily seal up the plant's stomata, preventing transpiration. When the sealant finally washes or wears off, the tree or shrub is usually able to survive on its own.

In regions of freezing winters, anti-transpirant sprays can also save the lives of broadleaf evergreens, such as rhododendrons and azaleas, when bright sunny days force open stomata at a time when roots may be quite dry under snow cover. The New York Times garden columnist Linda Yang recommends using anti-transpirant sprays as winter protection for evergreens grown in containers on balconies and roof gardens, since high-rise winds often dry out plants before a busy owner can get home to water.

Should the Circle Be Unbroken?

From sponge-like roots, liquids and nutrients flow up a plant's green stems and into the leaf-tips like juice through a straw. On bigger plants and thick-trunked trees, vascular flow may be better compared to a circle of straws, visible in cut wood as a thin green line between the dead bark and the cellulose-laden pith or heartwood. This green layer is called the *cambium*. Think of the cambium as the information highway for all woody plants—the center of vascular activity and plant growth.

During the growing season, cambial cells (see page 38) are always dividing. When one cell divides into two, one of the new cells remains a cambial cell, while the other changes and becomes either a *xylem* cell (to move water upward) or *phloem* cell (to move food downward for storage in fruits and roots). As the cells begin to crowd each other, they develop thicker cell walls; some push inward and die to become the woody center; some push outward, becoming part of the bark. This is how a tree trunk thickens over time. Some cambial cells change to become other specialized cells, eventually becoming shoots, leaves, or flower buds. This is how all plants grow.

Some gardeners call cambium the "lifeline" of the tree. Trees can handle slight wounds: cambial cells begin to transform, again, this time to create a woody callus that in time heals over the cut, just like a scab on a skinned knee. But slash the cambium on a large tree deep enough, and every bit of branch and trunk above that point will die, deprived of its vascular connection to the rest of the organism.

In America's early colonial period, settlers cleared the land in a way that saved their ax handles by simply *girdling* large trees—slicing the cambium all the way around the trunk. The tree would die in short order, and vegetables and tobacco could be planted over the dead tree roots. Today girdling is usually caused by a mistake or misfortune—the result, perhaps, of deer, rabbits, or bark-boring insects that have damaged the cambium by chewing all around a trunk. Sometimes the culprit is a gardener careless with a string-trimmer; weed-whacking scars are common tree-killers.

See for Yourself

1 Dig up the roots of a seedling tree, such as a maple or cedar. You'll find a large, thick central root, called a taproot, plus masses of medium-sized roots and, if you look closely enough, extremely fine, hairlike roots. These tiny roots are the feeder roots, thin enough to get between fine particles of soil. Thin cell walls make it easy for water and gases to pass into the root; then a process of vacuum suction moves the water and gases up into the main stem and beyond.

2 Cut a tree branch that is at least an inch thick to find the vascular network. Between the white or pale center of the heartwood and the thin shell of the bark, you'll see a green ring. That's the cambium, where the xylem (cells that send water upward in a plant) and phloem (cells that send food downward back to roots) live. Using a grafting knife, peel away the bark to examine the green part more completely. Notice how thin it is.

America's early farmers also knew how to manipulate the cambium to their advantage—using skills such as grafting fruit trees and pruning them to bear larger and heavier crops. Today's gardeners are once again learning these skills, thanks to groups such as North American Fruit Explorers (NAFEX). It is now possible to grow Thomas Jefferson's favorite variety of apple ('Esopus Spitzenberg') in your own backyard, from a tree grafted with fruiting wood that has been passed down through two centuries. Grafting is fun, and pruning is easy once you understand a few basic principles.

Apple

Up and Down

Plants don't have hearts, so in place of that pump they rely on basic physics to move water around. Since water molecules are cohesive, they tend to stick together and follow each other around. All it takes is a little bit of natural vacuum—i.e. the dryness of the air sucking out the first water molecule from the leaf stomata—to start lifting all the other water molecules through the plant's internal plumbing.

When water arrives from your garden hose and wets the soil, feeder roots absorb the groundwater, taking up dissolved solids of any minerals (iron, calcium, magnesium, etc.) from the soil or from fertilizers.

Water moves up the trunk stem and branches through specialized cells found just beneath the bark, visible as a green sheath called the cambium. Xylem cells move water and dissolved gases upward into each leaf, in much the same way that veins and arteries carry blood throughout our own bodies. Phloem cells carry back the dissolved sugars, starches, and gases that have been created within the leaves by the process of photosynthesis, taking these nutrients down to the roots, where they are stored.

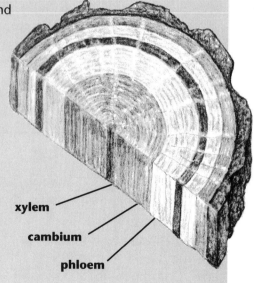

xylem

cambium

phloem

Experiments in Directed Growth

Many backyards have maple trees. If you've got one, try a series of summertime experiments that will show, in short order, how you can shape a tree with selective pruning. Maples make good subjects because their bark is thin and they grow fast.

Experiment One: The Weeping Maple

A maple in full leaf at the beginning of summer sends out long, flexible shoots; leaves appear first as pairs of buds that we see as raised, shiny bumps along the shoots. On maples, sets of buds alternate: if the first set grows vertically—up and down—the next set will sprout horizontally—right and left.

To make a weeping maple branch, pick a shoot on your maple and mark it with a ribbon. Next, decide which way you want that shoot to grow—up, down, right, left? Let's say you want it to grow downward, to become a weeping branch. Choose a plump set of vertical buds, and cut the shoot about a quarter-inch beyond these buds. Then, using your fingertips, pluck off the upward bud from the set, leaving the downward bud.

The shoot will now grow downward. Why? Because you've interrupted the vascular flow along the shoot. But the water flow in growing maples wants to stay constant, and the vascular activity has to go somewhere. The maple shoot quickly rouses its cambial cells to action on the remaining bud, and directs its growth to this area. The bud begins to elongate, and a new shoot is formed.

By repeatedly trimming to downward buds, you would, over time, develop a downward loop of growth. If you did this with all the shoots, you'd end up with a maple with a weeping silhouette, like something you'd see in a Japanese temple garden.

Experiment Two: Back-Budding

When you want to balance the look of a "one-sided" tree that has too many branches on one side, try this technique. It works with any fast-growing deciduous or evergreen tree. Elms, pines, junipers, and maples are good candidates.

Search the trunk of a small tree for the raised bumps that may indicate dormant buds; if you're lucky, you may find a bump that has sprouted a leaf or two. Choose a bud on the side of the tree that needs more branches.

With a sharp knife, gouge out a thin wedge of bark just above the bud. Make the cut at least 50 percent wider than the bud, and make sure the cut is deep enough to slice through and remove the green cambium. (You should see pale heartwood.)

Mark the spot with tree paint and keep an eye on it. What will happen is that the interrupted vascular flow will redirect itself and "force" the bud to grow out. In a few weeks (depending on the season and the weather), this dormant or weak bud will begin a new life as a side branch.

Experiment Three: Building Branches

The cambial cell's constant divisions can be observed with another pruning trick on fast-growing maple trees. This project is best done in mid-spring, when branches are putting out lots of long, soft shoots.

Choose a shoot and mark it with a ribbon. Then cut the shoot just above a pair of horizontal buds. Leave both buds intact, and odds are good that both will elongate into secondary shoots, creating a fork effect. Let each side of the fork grow out its own bud pairs, then cut each side back to a pair of horizontal buds.

These buds will form forks of their own. If you continue the process through the growing season, your original shoot will develop a fine branching structure, with leaves that fan out gracefully along a horizontal plane.

Use this technique on both evergreen and deciduous trees wherever you want them to grow branchy and full.

Pruning That Makes Sense

When you prune the large branches of street trees and other landscape trees, you're using the same technique for directing vascular growth that works for tiny maple shoots: If you shut off vascular flow in one direction, a tree will try its best to direct its energies to what remains. Careful pruning of landscape trees means simply realizing that most every cut is a binary choice. Cut here, and the plant will re-grow there.

Trees grow upward, usually with a central trunk called a *leader* that reaches high to become the apex or crown at maturity. If the central leader is removed or damaged, a side shoot from the trunk takes over and becomes the new leader.

Trees usually respond to wounds or cuts by sending out new shoots. In the hands of a bad tree trimmer, a really bad haircut for a landscape tree would be what's called "topping"—an ungainly attempt to shorten a tree's height by simply cutting the trunk down to the proper size.

Topping doesn't work on large, mature trees for several reasons: a severely pruned tree immediately begins to sprout wild new growth below the cuts. The new growth is thicker, but

As you prune, keep telling yourself it is for the good of the tree—if you're a convincing speaker it may help. If you still have doubts about the advantage of severe pruning, drive past a commercial orchard after it has been pruned and you'll probably be surprised at how much wood they cut out.
—Lewis Hill, *Fruits and Berries for the Home Garden*

Pruning Questions and Answers

When is the best time to prune?

Deciduous trees, including fruit trees such as apple and pear, are usually pruned during the dormant period, when they have lost all or most of their leaves, and vascular activity slows considerably.

Evergreen trees, including semi-tropical citrus, can be pruned at any time of year. Frost-tender species, such as citrus and podocarpus, should not be pruned as winter approaches: a snap freeze may kill off branches with exposed cuts.

Flowering shrubs and vines, such as lilac, forsythia, crape myrtle, camellia, wisteria, jasmine, azalea, and rhododendron, are ideally pruned shortly after the flowers have faded. Because it takes nearly a year for flower buds to form on these plants, pruning done earlier or later will be at the expense of your floral display.

Roses are usually pruned during their dormant, leafless state, from late fall to early spring. In mild-winter climates, roses don't always go dormant easily. To give them a proper rest, stop cutting blossoms after Halloween, and let the bush form fruits (called hips). After this, the leaves will fall, and any left can be pulled off the bush by hand.

Summer pruning is a technique to thin out too-thick branches on deciduous trees and shrubs that make heavy growth in summer: this includes grape and wisteria vines, and most fruits in the vigorous *Prunus* family, which includes peaches, apricots, plums, and the decorative purple-leaf plum. In this case, the careful gardener will remove ungainly shoots that have no fruit on them (called *watersprouts*). The job is usually done when a tree has leafed out fully (June to August) and at least a month before autumn leaf drop.

Do fruit trees need special pruning care?

Trees, bushes, and vines that produce edible fruits usually do so on or near the tips of their new growth.

Since different fruits differ in their needs, it's best to consult a good pruning handbook, such as Lewis Hill's 1979 classic, *Pruning Simplified*, or his later book, *Fruits and Berries for the Home Garden* (see Sources) if you do this job yourself.

Any advice on trimming hedges?

If you're planning a special barbecue for July 4, don't wait any later than Memorial Day to do a big trim—otherwise, the plant may not recover in time to look nice. To keep hedges bushy and full, trim lightly and frequently over the growing period. Major surgery should be left until the next dormant winter season.

Do I need to seal pruning cuts?

Extensive research from the University of California suggests that latex, tar, and other tree sealants—or using nothing at all—makes absolutely no difference to the rate at which a pruning wound will heal.

Whip Graft

New Trees from Old: A Whip Graft for Apple Trees

This is an easy grafting project for beginners. If you do the job during the dormant winter season, apple tree grafts will "take" in a single year.

You'll need the following supplies (see Sources):
- Apple rootstock or backyard apple tree
- Apple scionwood sticks of an interesting variety (cut fresh or store in the refrigerator until used)
- Grafting knife
- Cut rubber bands
- Grafting sealant, such as wax, parafilm, or Doc Farwell's Seal & Heal latex grafting seal

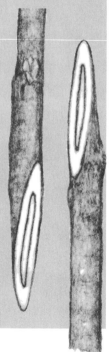

1 To get a good cambium match, the diameter of the scionwood should equal the diameter of the rootstock branch you want to graft to. Slightly thicker than a pencil is a good size for apple scions; bring the scionwood over to the rootstock tree and search till you find a branch that is of similar girth.

2 Keep the scionwood moist in a cup of water. (The best grafter I know sticks the cut end in his mouth until he is ready for it.)

3 Cut the tip off the rootstock branch with the grafting knife, making an elongated, diagonal cut. The object here is to expose the largest area of green cambium.

4 Cut the scionwood on the same diagonal slant, so the two pieces will fit together as if they were one branch. If the pieces don't match up the first time, keep cutting and trimming. Practice makes perfect; just be careful with that knife!

5 Press the scionwood and the rootstock together, cambium to cambium. Tie them securely together with a cut rubber band.

6 Apply a grafting sealant to completely cover the cut areas and the rubber band. This will prevent the graft from drying out. Some grafters also wrap a paper bag loosely around the graft to shield it from drying wind and sun.

7 Tie a ribbon or metal tag to the grafted branch so you will remember where it is. By summer, you should see new leaves growing from the scionwood; this is the sign that the cambium has fused successfully. Slit the rubber band and/or sealant at the end of the summer so the branch can continue to thicken normally.

weak and wild, and it soon will have to be pruned again, though the natural and beautiful line of the tree is destroyed forever.

A better technique for reducing the height of a mature tree is called *crown reduction*. Instead of simply cutting to a shorter height, the thoughtful tree trimmer examines the leader to find a nearly parallel, upward-thrusting branch at a lower level. When the trunk is cut just above this branch, the lower branch will become the new leader. The tree trimmer also examines main branches along the trunk and prunes them down to the point where there's a thinner, smaller, and parallel side branch. This technique preserves the natural lines of a tree while reducing the overall height in a graceful way. (For more about crown reduction, see Sources.)

For many decades, professional arborists (tree trimmers to you) advocated removing large branches by cutting them flush to the trunk. New research has changed this advice: it's now suggested that large branches should be cut back only to the *branch collar*— the little ring of raised bark where the branch is joined to the trunk. Err on the side of length, experts say: the branch collar is apparently an active center for cambial cells, and callusing of the cut occurs much faster if the branch collar remains undamaged.

More Fun with Cambium—Grafting

Knowing about vascular flow also makes it possible for you to graft one tree or shrub onto another. Roses and fruit trees are routinely grafted by surgically attaching a living bit of a wanted variety (called the *scion*) to an existing tree (called the *rootstock*).

In the grafting game, you try to align cambium from the scion with cambium from the rootstock tree, so the vascular flow runs smoothly from the old wood to the new. There are many

Apple tree

techniques for attaching grafts; some involve merely slitting the bark of the rootstock tree and inserting a live bud that, with luck, will merge with the cambium. (This is sometimes called *shield budding* or *T-budding*.)

Most plants within the same species or genus can be joined: you can graft five different varieties of apple (*Malus Malus*) onto the branches of a single tree; you can graft five different colors of roses (*Rosa*) onto a single bush. A family resemblance among all *Prunus* (apricots, almonds, cherries, peaches, nectarines, plums, apriums, and plumcots) lets you graft all these interchangeably. But you can't graft an apple (*Malus*) onto a pear (*Pyrus*) because these two genera are so genetically different.

Some Last Words on Water

Good water management is not just important to plant health, it's important to maintain a healthy planet. What pours out of our garden hoses drains far into the ground, affecting tiny organisms that live in the soil and mingling with groundwater, underground streams, and drainage systems, eventually flowing into rivers, lakes, and oceans near where we live. To keep this water clean, gardeners need to consider the amount of mineral or chemical additions they add via fertilizers and soil amendments, since what plants don't use invariably will wind up downstream, and perhaps back in your drinking glass.

Learning to conserve clean water should be part of every gardener's life. In my kitchen I keep a few empty plastic gallon milk jugs near the sink, and these get filled up while I'm waiting for the tap water to get hot or cold. What's collected gets used to water window boxes and potted plants. In the garden I have an

"To use drip irrigation without using mulch would be like wearing pants without a belt—you get most of the effect, but without that last measure of protection."

—Robert Kourik,
Drip Irrigation for Every Landscape and All Climates

automatic timer on my sprinkler system, which turns on at sunrise, the best time to irrigate—no need to get up at the crack of dawn to water the yard. Many gardeners I know have installed drip irrigation systems, which deliver water to root zones only and can be highly efficient. (For more on drip irrigation, see Sources.)

Mulching is a once-a-year operation I highly recommend in all gardens. If you set down a three- to four-inch layer of fluffy compost, shredded bark, cocoa hulls, or other mulch between your plants, this will cut down on weeds that "steal" water, and it will do a great deal to slow down the rate that water evaporates out of the ground. Mulching should be done in early spring, while the soil is still moist and cool—don't wait until the summer sun has baked the ground and your earthworms have fled—and rake off the mulch in late fall so it will not be a haven for rodents or overwintering spores of pesky fungus.

If you live where droughts are common, water-saving techniques like these might make a difference as to whether or not you can have a garden at all.

I can assure the uninformed that a proud pounding of the heart greets the moment when the dormant bud of the scion, which had been sleeping on the foreign stem, wakes, turns green, asserts its paradoxical nature, and enjoins the eglantine with the rose, the plum tree with its peach, its nectarine.

—Colette, *Flowers and Fruit*

Diagnosing Common Plant Diseases

A Vascular Perspective

Here are some common garden ailments that an astute gardener can diagnose by examining a plant's leaves.

• Root Rot or Crown Rot *(Phytophthora)*

Symptoms: Plant begins browning at the leaf-tips, then discoloration and wilted leaves begin to spread from the branches on down. Common to drought-hardy species, such as juniper and California oak, also to strawberries and container houseplants.

Cause: Root rots are caused by a number of slime mold fungi that live in soil. Constantly damp or water-saturated soils encourage them; they kill by clogging up the conducting tissue from roots to stems.

If you suspect root rot, dig up a bit of root and cut it on the diagonal. A healthy root will be white and clean-looking on the inside; a root attacked by rot will have brownish discoloration seen as stripes or lesions. Root rot can be fatal, and there is no chemical cure.

Treatment: Let the soil dry out thoroughly; stop irrigating drought-hardy species and see if they will recover. If the plant is in a pot, trim off affected roots and transplant to clean potting soil that is a well-drained mix.

• Sunburn

Symptoms: Irregular white or brown discoloration in blotches between the veins of leaves and on leaves, indicating loss of green chlorophyll. Cracked bark, but only on one side of the tree.

Cause: Inadequate watering and too-dry soil cause this scorching. On tender, leafed vegetables and houseplants, sunburn may also be caused by overhead watering during bright, hot weather; water droplets act as magnifying glasses and literally burn the chlorophyll off the leaves, leaving sunken white or brown spots.

Treatment: Plants will recover with irrigation, and the brown leaves can be cut off when you see new growth. To prevent sunburn on garden trees, paint trunks with white paint during their early years.

• Leaf-Tip Burns (Salt Burn)

Symptoms: Leaves look wilted; tips are blackened and shriveled. Common to many container plants.

Cause: Overdose of fertilizer. Common in container specimens when chemical salts build up in the restricted soil of a pot that is watered lightly instead of deeply.

Treatment: Flushing the soil thoroughly whenever you water a potted plant is usually enough to prevent salt buildup. In the garden, accidental overdose from salt spread on roads and driveways during icy weather may be treated by irrigating deeply to flush salts away.

(Plants grown near the ocean are sometimes affected by salt spray; choosing plants native to seashore ecosystems can be a good solution for both roadway plantings and seaside vacation homes.)

• Chlorosis

Symptoms: Leaves are healthy but yellow or partly yellow. Common in evergreens, lawns, and houseplants.

Cause: Too much watering. Excess irrigation can flush out nutrients required to create healthy chlorophyll—particularly nitrogen, iron, and magnesium. Outdoors, an especially rainy winter or spring can leach out nutrients in garden beds and cause evergreens to turn sickly pale.

Treatment: This problem can be treated quickly by applying a mineral-rich fertilizer to soil. Chelated iron, in powder or liquid form, will green up yellowed lawns in a few days.

Fusarium, Verticillum Wilt

Symptoms: Leaves turn yellow, wilt, then turn brown and die. Common to tomatoes and other vegetables.

Cause: Wilts are caused by fungi in soil, which attack roots that may be damaged or broken by overzealous tilling or clumsy transplanting. Cutting a stem into cross section often shows brownish or yellowish stripes or lesions.

Treatment: Affected plants should be torn up and discarded (into the trash, not the compost pile). Since wilt spores can live for up to five years in soil, crop rotation and the use of resistant varieties can keep wilts under control in the vegetable bed. To treat a larger plant such as grapevine, cut off infected leaves and branches, apply a fungicide, and hope for the best.

Powdery Mildew, Downy Mildew

Symptoms: Whitish or grey filmy bloom on leaves and flowers; flowers are misshapen; fruits have crackled skins.

Cause: Like the mildew that forms on your bathroom shower wall, mildews of the plant world are airborne fungus that attach and grow in warm, moist spots. Their fuzzy fruiting bodies clog up leaf stomata and interfere with transpiration, which weakens the plant.

Treatment: Plant mildews are host-specific: the fungus that grows on a rose is not necessarily the same organism that prefers a squash leaf or zinnia blossom. Fortunately, all mildews can be eradicated and even prevented by applying a chemical or organic garden fungicide at biweekly intervals during the growing season.

Know Thy Dirt

3 The Physics and Chemistry of Soils

ONE CHRISTMAS WHEN I was a little girl we had a live fir tree to decorate. After the holidays, we planted it in the ground alongside our garage. As the years went by, we kids grew bigger, but the tree did not. It remained a green but stunted thing, just a few feet high, even though we tilled the soil and planted flowers around it and gave the little tree fertilizer and water.

In the summer before I entered high school, the dwarf tree put on a tremendous growth spurt. Doubling in width and bushiness, its pointy crown soared six feet in a season to reach the top of the garage wall.

What had changed? That was the summer Dad had decided to replaster and whitewash the garage walls, using a mixture of white cement and hydrated lime. When summer rains pelted the garage wall, lime residue dripped down into the soil around the Christmas tree. Lime is a powerful soil additive. This altered the chemical structure of the soil near the tree's roots in a manner that caused the tree to suddenly thrive.

It was a dramatic lesson in soil chemistry, and how having different soil ingredients can make a big difference in how plants grow.

What's in Soil?

Soil is the crust of our planet, earth. On undisturbed land, it appears as fragmented particles from the rock of each continent's landmass, eroded by wind and water. On its topmost layer, the particles may be fine as dust; where trees and other plants grow vigorously, their organic remains may be a part of this top layer. We call this fluffy, rich mix topsoil. Organic materials usually give topsoil a dark color.

The fragments get larger and larger as you go lower down into layers of subsoil. Subsoils are typically lighter in color, since the only organic material present is what has sifted down through cracks from the topsoil. Subsoil may be densely compacted; mineral deposits from accumulating water percolating down to this layer may mass themselves into impermeable strata known as hardpan or *caliche*. Below the subsoil layer are even bigger chunks of base rock, broken only by cataclysmic actions, such as seismic heavings. Below this is the base rock, also called bedrock.

> *"Nature is such a subtle chemist that one never knows what she is about, or what surprises she may have in store for us."*
>
> —Gertrude Jekyll, Wood and Garden

The gardener is really only interested in topsoil. This is where the roots of plants will go in search of the molecules that are their building blocks: carbon, oxygen, nitrogen, and minerals such as the iron needed to create chlorophyll for green leaves, and the phosphorus needed to stimulate reproduction via flowers and seeds. Organic material provides some carbon and most of the nitrogen; water and air percolating between soil particles supply oxygen; the soil particles themselves may supply the needed minerals.

"Most of my bearded iris have suffered much from theory."

—Elizabeth Lawrence, A Southern Garden

In desert or tundra areas, where organic material is scarce, topsoil may only be a fraction of an inch thick. In peat bogs, jungles, and dense old forests, the topsoil layer may be several feet deep. In your garden, you can create more topsoil by systematically adding more soil particles and extra decomposing plant materials in the form of compost or chipped bark. Minerals can be added as well. We call this process *fertilizing*— our attempts to make the soil more fertile, so it will yield more apples, cucumbers, lilies, and lilacs.

Fertilizers and Soil Structure

A big mistake is thinking that fertilizing plants is like feeding an animal—give the plant "food" by applying fertilizer and the roots will "eat" the food. What really happens is an unconscious and seemingly automatic exchange of molecules between the roots and the soil.

As with true love, the secret of good garden soil is simply a matter of chemistry. Roots, like leaves, transpire gases and liquids during periods of active growth. As roots spread through soil, their growth process releases hydrogen ions into nearby soil.

Sow bug

An *ion* is an atom or group of atoms that has lost or gained one or more negatively charged electrons, which gives the ion a positive or negative charge. Hydrogen ions are hydrogen atoms that have lost one electron, so that they are positively charged. The positively charged hydrogen ions produced by roots create a slightly acidic soil condition just around the root zone. This acidity helps break down soil particles and release nutrients that plants can absorb.

Nitrogen fertilizers, for example, break down in the soil into absorbable negatively charged ions made up of nitrogen and oxygen. (The chemical formula for this ion is NO_3^-, which means the ion includes one nitrogen atom and three oxygen atoms and it has a negative charge.) Phosphorus is absorbed as another negatively charged ion, $PO_4^=$, which includes one phosphorus atom and four oxygen atoms.

In the root zone, an ion exchange occurs. Some of the fertilizer's ions detach from the surrounding soil and are simply carried by the flow of water through cell membranes into the plant's root cells. The entry of dissolved nutrients via water—achieved through the suction action discussed in Chapter 2—should underscore the importance of watering well before applying any fertilizer. There must be enough water present in the root zone for what's called *solvent drag* to pull those mineral ions in.

Centipede

Other ions are actively transported across cell membranes. Scientists are still trying to figure out exactly how this process works, but it seems to involve some sort of ion swap, where one type of ion (say, a negative phosphorus ion) is transported into the root while another (perhaps a positive hydrogen ion) is transported out.

So fertilizers are absorbed, but not in great draughts; the job is done molecule by molecule, sometimes ion by ion, at a subatomic level in the root zone. Each plant's roots choose ions needed by its particular species at particular times of the year. That's why roses, lawn grass, tomatoes, and fir trees all seem to need different kinds of fertilizers, which can often be found packaged separately at garden centers under the name "lawn food," "rose food," etc. But it isn't enough to simply dump on a bunch of the appropriate fertilizer and hope for the best.

An earth-science approach to gardening looks first to see if conditions are favorable for an active ion exchange. A loose and fluffy soil texture is desirable, since a free flow of water and gases in the root zone would provide the widest opportunity for ions to meet. Topsoil compacted by pelting rains or trampling feet is less effective, which is why gardeners often grow vegetables in raised beds or use a technique called *double digging* to loosen up the ground before they begin planting.

Room to Grow?

The air spaces between soil particles vary with the size of the particles. Soils are usually characterized by the size of their particles, categorized as either sand, silt, or clay.

Sandy soils have gritty, rough textures. The soil particles are usually large enough to see with the naked eye; the spaces between particles are quite large as well.

Silty soils have a finer grain, and are usually found in alluvial deposits and former stream beds.

Clay soil particles are the smallest. Air spaces between clay particles are correspondingly minute, which is why clay soils often drain poorly. Yet some mineral-rich clay soils also permit the highest levels of ion exchange. Clay particles chemically bind or trap nutrient molecules so they are not washed away, while the large number of particles multiplies the surfaces from which molecules can project their ions, compensating for the limited space to do the transaction. In the U.S., agricultural states, such as Georgia, are renowned for their fertile clay soils.

The ideal garden soil has a little of each: some sand, for good drainage; some silt, for mineral richness; some clay, to boost ionic exchange capacity and retain water during dry spells. And to keep the mineral elements loose and fluffy, garden soil should also have some *humus,* the gardener's term for decayed and decomposed organic materials. If your garden is short of humus, a good substitute is *compost,* which can be made by decomposing plant debris. (See page 59.)

Double Digging

Before double digging, mark out and water a site in your garden—three to five feet wide and twenty feet long is a good size to start with—and loosen the top one-foot layer of soil with a spading fork. Remove any weeds, and cover the bed with an inch of finished compost.

Place a board across the bed to spread your weight and avoid compacting the loosened soil. Then use a spade to dig a trench about one-foot deep and one-foot wide across the width of the bed. Soil from this trench can be placed in a wheelbarrow and saved—you'll use it when you prepare the bed's last section.

Next, use a spade or digging fork to loosen the soil in the bottom of the trench, being careful to keep the soil layers intact.

This next step is where you do your best to imitate a landslide: after sliding the digging board back to reveal another foot or so of the bed, 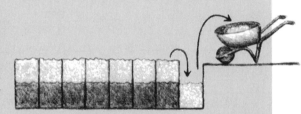 use a spade to dig out the top one-foot by one-foot section of a second trench, then heave that soil forward into the first trench. Again, try to avoid mixing the soil layers. Most productive soil organisms live in the top several inches of soil. These organisms help break down the compost and other organic matter being added to the bed. Buried too deep, the soil life can suffocate.

Repeat the steps listed above until you reach the end of the bed, and fill in the last trench with the soil you set aside earlier. Finally, work fertilizers, such as bone meal or kelp meal, or additional compost into the top several inches of soil with a spading fork, and rake the bed's surface smooth.

—Martha Brown,
"Dig Deep to Make Your Garden Bloom,"
Exploring magazine, Spring 1996

> t goes without saying that if the soil is heavy, it should be made lighter. ... If you have no manure, than you must use peat moss or thoroughly rotted leaves or whatever other humus you can find. It is fairly imbecile to say that one has no peat, no sand, no manure, no compost. Get some. You should have had it years ago, and unless you want to settle for pokeweed, you need to produce a soil that is brown and crumbly.
>
> —Henry Mitchell, *One Man's Garden*

Living Organisms in Soil

In average soil, 1 to 8 percent of the material present is organic material. That is "organic" in its strictest sense: either living examples of carbon-based life-forms, such as soil bacteria or earthworms, or the dead remains of these or other once-living creatures. Soil life doesn't take up a lot of space, but it's surprisingly abundant: a teaspoon of earth from your backyard may contain billions of microscopic organisms.

These living creatures play an important role in the life of your garden plants. Their biggest job is decomposing, breaking down introduced elements—such as fallen autumn leaves, rusty nails, soil clods, and the dead bodies of other animals—into their elemental forms (nitrogen, calcium, oxygen, etc.) so plants can absorb these nutrients.

Soil bacteria, actinomycetes, and underground fungi are the microscopic clean-up crew for your soil. Bacteria are single-celled organisms that convert sugars to carbons, or draw oxygen and nitrogen into their bodies. Actinomycetes are similar to bacteria; you may not recognize their name, but all good gardeners recognize their smell. The rich and pleasant odor of freshly turned earth, so redolent in the cool seasons of spring and fall, comes from chemical gases released by actinomycetes as they transform dead plant material into usable humus.

Soil fungi (like all fungi) are related to mushrooms. When you turn over fresh earth on a forest floor, you will often see the white

Making Compost

First, you need a compost bin. I bought a prefabricated bin from the local garden supply store, but you can also build one out of plywood and chicken wire, dig a pit in the ground, or even use a plastic garbage bag or garbage can with holes poked in the bottom, top, and sides for drainage and aeration. (If you're using a plastic bag, you should leave it open every other day to let air in.)

A simple recipe for making compost is to layer one-third dry (brown) material, one-third green vegetation, and one-third soil. (Soil provides a starter supply of microbes to get the compost going.)

After that, almost everything goes: kitchen waste, grass clippings, dry leaves, dead plants, coffee grounds, shredded newspaper, even clothes-dryer lint and pet hair. When adding ingredients, you should strive for a 25-to-1 ratio of carbon sources (brown stuff like dead leaves and newspaper) to nitrogen sources (green stuff like grass clippings and foliage). Green materials contain carbon as well as nitrogen, but dead leaves and other brown materials are largely devoid of nitrogen.

Cattle, horse, and chicken manure are excellent compost ingredients if you can get them. Avoid pet and human wastes (they may spread diseases), and meat, fat, and weed seeds or diseased plants.

To give your compost a head start, chop up or shred materials before adding them to the bin. Periodically sprinkle the compost with water so it stays moist but not soggy—it should be about as damp as a moist sponge. With a pitchfork or spade, turn or stir the compost to introduce air. Compost that is too wet or isn't exposed to enough oxygen can turn stinky. An easy cure for this is to open the bin and let the compost dry out for a few days.

—Mary K. Miller,
"Making Compost,"
Exploring magazine, Spring 1996

strands of their threadlike *mycelium,* extended networks of cells that produce digestive enzymes that break down woody plant material into a food source for the fungi and for nearby plants. Fungi are particularly good at breaking down phosphates into usable phosphorus. Some current research has involved "seeding" fungi into root zones to act as a fertilizer stimulant.

Bacteria, actinomycetes, and fungi retain the minerals and molecules they have absorbed within themselves until they die or get eaten by larger organisms. This is one of the reasons why organically rich soils are said to be more fertile; the microbes act as holding tanks for useful soil nutrients, released over time (think of the tiny time capsules found in slow-release fertilizers) as they are gobbled up in the course of a growing year.

The predators that do the gobbling include soil mites, sow bugs, nematodes, and springtails—tiny creatures most gardeners consider to be "insect pests." Sow bugs (also called pill bugs) roll into a ball when touched, and often get a bad rap for destroying fruits such as strawberries, when the gardener finds them curled up asleep in a hole in a berry that was really chewed up by something else. Sow bugs really prefer decaying vegetation to fresh stuff. It's true some nematode species are harmful to plants (especially tomatoes), but most of these tiny snake-like soil dwellers are helpful allies because they eat up soil bacteria and then excrete or pass on their nutrients.

Every gardener likes to see earthworms. They eat up tons of soil microbes and decaying vegetation as they tunnel through the ground, then pass on the nutrients

Earthworm

through their excretions, called castings. Worms also perform a useful physical task. Their underground tunnels open up the soil to air and water, reversing the effects of soil compaction and making pathways for plant roots to penetrate.

It is not absolutely essential to have underground guests—or even soil—to grow good plants. Lettuces grown hydroponically, with their roots suspended in liquid, or rare conifers grown in tubs of sterile gravel will thrive just as well if dosed with plant nutrients in the equally sterile form of a water-soluble fertilizer, such as Miracle-Gro. "Soil-less" potting mixes are often used for starting flower and vegetable seeds, because the sterile medium assures there are no harmful soil disease bacteria or nematodes that would kill off sprouting seedlings. In this case, too, nutrients need to be introduced through the application of a fertilizer as the seedlings grow.

The Mystery of pH

Fertilizer may reach a plant's roots, but the roots' ability to absorb nutrients through ion exchange will vary depending on the chemical conditions in the material surrounding the roots. The availability of positively charged hydrogen ions for ion exchange can be adjusted and may be monitored through a measurement known as *pH*.

Gardeners often loosely describe the term pH as "percentage of hydrogen." But, the numbers run backwards: a pH number technically measures a *negative* logarithm of hydrogen ion concentration, on a set scale of 1 to 14. So, the higher the pH number is, the fewer hydrogen ions there are. To make matters even more confusing, the scale is exponential, like the Richter scale for earthquakes. So a pH of 14 (practically unheard of in garden soils) would indicate an extremely low concentration of hydrogen ions, so low it would not be able to sustain any plant life at all. A pH of 7 is considered neutral; we call soils with a lower pH number acid soils. Soils with a higher pH number are called sweet or alkaline or (in England) chalky soils.

How acidic is an acid soil? Soil with a pH level of 3 has about the same level of puckery acidity as a grapefruit. Soil with a pH level of 8 has roughly the same alkalinity as a teaspoon of baking soda. These levels are too extreme for most garden plants, which reside comfortably in a pH range between 5 and 7.

Individual plant species are just as picky about their pH needs as they are about sun and water. Garden rosebushes and most sun-

See for Yourself

1 Is your soil sandy, silty, or mostly clay? Try this easy test: Grab a fistful of average soil and squeeze it tightly. If it crumbles loosely and quickly, you've got a mostly sandy garden bed. If it crumbles slowly, you probably have a good mixture of sand, silt, and perhaps some clay. If the soil sticks together in a clump that retains the imprints of your fingers—then you've got a clay soil, for better or worse.

2 If you would like to determine the exact proportions of sand, silt, and clay in your soil, you can get a pretty accurate reading using the following household items:

- Clear glass quart jar with lid (such as a mayonnaise jar)
- One teaspoon water softener (Calgon works the best)
- Stopwatch or clock with a second hand
- Marker or grease pencil

Fill the quart jar ⅔ full of water, and stir in the teaspoon of water softener. Fill the jar nearly to the top with a sample of soil from your garden.

Screw on the lid and shake the jar vigorously. The water softener will *deflocculate,* or separate, the particles, and suspend them temporarily in the water solution.

Set the jar down on a flat surface and start your stopwatch. After 20 seconds, you will see a dark line forming near the bottom of the jar. These are the large particles of sand in your soil, which are settling first because they are heaviest. Without disturbing the jar or its contents, mark this line on the outside of the jar with the marker or grease pencil.

Keep an eye on the watch and wait two minutes more. A second line of settling material will begin to form. This is the silt within your soil. Mark it with the marker.

A third visible stratum of material will begin to settle down, although depending on the amount of clay in your soil it may take anywhere from overnight to a week for all the clay particles to settle. This is because clay particles are particularly lightweight and susceptible to deflocculation. (Artistic potters sometimes use Calgon in their clay mixes to create unusual textures in ceramics.)

When the water looks clear, the clay is settled, and you should see three very distinct bands of soil deposits within the jar. These are the proportions of sand, silt, and clay in your soil sample. Rough material sitting on the top of the layers, or floating near the surface of the water in the jar, is organic matter in your soil mix.

3 Get a soil testing kit, such as rapitest or LaMotte, from your local garden supply store. Using testing tablets and small samples of soil in test tubes included in the kits, you can get a reading on soil factors, such as pH and the amount of available nitrogen, potassium, and phosphorus in your sample. Mixing the sample with distilled water (not tap water) produces the most accurate results in a home test, since tap water may contain minerals and can be variable in its pH factor. You'll find distilled water in the mineral-water aisle in your grocery store. For a small fee, mail-order soil testing labs can provide the most detailed analysis of soil samples. (See Sources.)

Shifting Soil pH

he effects of pH shift are easiest to see in the controlled environment of a potted plant. Chase Rosade, an American bonsai master from New Hope, Pennsylvania, has been sharing this experiment across the U.S. for many years. He says it works best with conifers, such as the junipers, pines, and firs grown in small pots for bonsai work. The observation can also be made in potted maples and azaleas.

Traditionally, these plants are grown in acid soil mixes and/or routinely fed with acid-shifting fertilizers, such as Mir-Acid or cakes made of cottonseed meal. If you have a potted conifer, Rosade suggests sprinkling one tablespoon (for large containers, two) of hydrated lime into the soil in very early spring. Since hydrated lime is quick-acting, but will also leach away quickly in daily watering, the soil will have a brief but significant shift in pH. The shift releases previously bound-up ions into forms more accessible to the plant, and may be sufficient to prompt a quickie ion exchange with soil minerals that have been previously unavailable. This usually results in new vigor for the potted plant.

Try it yourself—but don't be tempted to do it more than once a year. Measuring soil pH before and after the treatment, and monitoring the results over the course of the growing season, can help you fine-tune the technique.

loving plants, such as English lavender and lawn grass, prefer a pH around 7, and can tolerate even more alkalinity. Woodland plants and shade lovers, such as blueberry bushes, azaleas, and rhododendrons, prefer acid soils and a pH around 5.5.

The important thing for a gardener to remember is that if a soil pH does not match a plant's needs, ion exchange and ionic absorption will be inhibited. When this is the case, the plant is physically and chemically unable to take up sufficient soil nutrients. In fact, it would not matter how often or how much you fertilized that plant: if the pH is off, the roots just won't absorb the fertilizer. This would be like giving a hungry man a bowl of tasty tomato soup—and then expecting him to eat it using a fork, not a spoon.

You can determine the pH level of any soil with a soil test kit (available at any garden center) or a pH meter probe. Soil pH can be easily adjusted, too. The addition of agricultural soil additives, such as powdered sulfur, chelated iron, or aluminum sulfate in small amounts, will alter the chemical structure of soil to lower the pH—or, as gardeners say—it will make the soil more acidic. To raise the pH level, and make the soil more alkaline, the chemical structure of the soil can be altered by adding ground limestone, dolomite lime, crushed oyster shells, or hydrated lime.

This, of course, is what caused the dramatic leap of growth in my family's Christmas tree. Hydrated lime in the whitewash on the garage wall seeped into the ground and began to alter the pH of the native soil—the generally acid soil of New Jersey that does such a great job growing tomatoes, peaches, and the pretty, pink-flowered mountain laurel. The Christmas fir had languished in this acid environment, and only thrived when the soil's pH was raised to a level it preferred.

Lavender

Deeper Secrets of pH

To complicate matters further, a plant's roots will do an ion exchange with certain mineral molecules only at a certain pH level. When this level is adjusted up or down, the plant may gain or lose the ability to absorb that mineral. Suddenly, different nutrients will be absorbed, and the growth of the plant can be manipulated in some interesting ways.

A common trick involves hydrangea: in acid soils, both the common mophead hortensia and the delicate lacecap (both *Hydrangea macrophylla*) will produce flowers of a sparkling deep blue; the flower pigments are the result of mineral intake of aluminum and iron. In alkaline soils, hydrangeas are unable to absorb these minerals, so they lose the ability to manufacture blue pigment, and will bloom in shades of dull red or pink. Adding aluminum sulfate to acidify the soil around the roots each spring brings back the beautiful blue color.

A more practical application of this information comes via the Cornell University Cooperative Extension, which maintains an information center for urban gardeners in downtown Manhattan, not far from Herald Square. Horticulture advisors there were worried about New Yorkers who grow fruits and vegetables in community gardens, backyards, and rooftops—sites not far from highly trafficked streets and highways, where the air and soils are polluted with chemicals from auto emissions.

The scientists tested city-grown produce and discovered significant amounts of cadmium, lead, and other heavy metals in fruits, such as tomatoes, and in leafy greens, like lettuce. Some toxic residue came from air pollution and could be washed away. (Cornell recommends rinsing city crops in a solution of one part vinegar to nine parts water before cooking or serving.) But some of the lead did enter the plants via the roots, so it couldn't be washed away.

The scientists explored further and discovered that lead and heavy metals are absorbed through ion exchange by vegetable plant roots when soils in garden beds and containers have a pH level of 6.5 or higher.

Cornell recommends that city gardeners worried about lead and metals in soils adjust their soil pH to more acidic conditions, to prevent or at least slow the absorption of toxic minerals into food plants. Acidic soils are more easily leached of metal ions, so routine irrigation might suffice to flush the toxins after the pH adjustment. If you garden near a busy city street, or grow your tomatoes in rows next to an active driveway, use these techniques and consider having your garden soil tested for heavy metals, especially lead.

Adjusting Soils to Suit Garden Plants

As we learned in Chapter 1, garden plants from one side of the globe can easily be grown on the other side of the world once you understand what their environmental needs are in terms of sunshine, water, and soils. Over the centuries, farmers and gardeners learned that growing the same plants repeatedly in the same soil seemed to drain the earth of nutrients; they also learned that these nutrients could be replaced by digging-in certain soil additives year to year. These soil-amending recipes have been passed down like secret handshakes, gathered in some of the earliest printed books, and today zip back and forth in cyberspace, as gardeners converse and trade information through the Internet and the World Wide Web.

Who was it that first dared to suggest that cow manure might be a useful addition to a field of oats or hay? In some parts of the world it is still acceptable to spread human excrement ("night soil") around vegetable beds, where predominately vegetarian diets minimize the risk of parasite infections. (This may provide one of the ultimate examples of true closed-circle recycling.) In our own country, mined and manufactured fertilizers share shelf space in garden centers and feed stores with exotic "natural" additives, such as bat guano culled directly from caves in Argentina, and powders made from kelps and seaweed.

What, you might ask, is supposed to be in this stuff? All plants, from algae to mighty redwoods, rely on three main elements for growth, and they are the main numbers you can read on a fertilizer bag: N-P-K, for nitrogen, phosphorus, and potassium. Certain other minerals are also used by plants to develop cells and reproduce.

Nitrogen promotes leafy green growth in all plants, and is essential in creating plant proteins—including the food proteins found in edible crops, such as beans and corn. Nitrogen is present in organically rich soils, available in fixed forms derived from animal wastes and ureas found in urine. It is also captured from the air by bacteria (called "nitrogen-fixing bacteria"), which attach themselves to the roots of leguminous (bean family) plants. As a result, these plants tend to contain high levels of nitrogen in their leaves, stems, and seeds. Animal manures and by-products, fish wastes, manufactured ammonium sulfate, and nitrogen-rich plant debris (such as cottonseed meal, alfalfa, or clover) are good sources of nitrogen.

Phosphorus provides strength to plant tissues, and promotes strong root growth and the development of flowers and fruits. Phosphorus is a mineral occurring naturally in most soils. Powdered rock phosphates or manufactured superphosphates are common soil additives, as is bone meal derived from fish or mammal sources.

Potassium is a mineral that helps regulate water flow in every single plant cell; it also affects fruiting and flowering. It may be present in soils or arrive as a residue found in wood ashes after

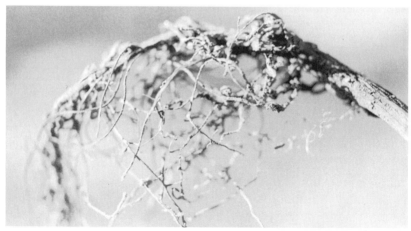

Fava bean root

Grow Your Own Nitrogen

Plants in the pea and bean family *(Leguminosae)* attract a special soil bacteria that captures nitrogen from the air itself. The bacteria live in colonies along bean plant roots, which siphon off the nitrogen in exchange for hosting the bacteria. As these bacteria multiply, it is sometimes possible to see the accumulated nitrogen stores as visible white nodules when the roots are pulled out of the soil. Peas, clover, soybeans, alfalfa, tepary beans, and vetch are some nitrogen-fixers that have been used in crop rotations to restore soil fertility since the days of ancient Rome and the Incan empire. Planting an acre of land with alfalfa can return 300 pounds of usable nitrogen to the soil.

If you want to see some really big nodules, plant fava beans (*Vicia Faba*), sometimes called broad beans or horse beans. A European edible with a long history, favas are a cool-season crop to be planted in the early spring (in mild winter climates, plant them in mid-autumn). Fava vines are vigorous and thick, and need support, such as a pea fence or pea sticks. White flowers that appear on the vines give off a honey-sweet fragrance at night.

Pull up a fava plant at the flower stage and you'll easily see bright white balls—nitrogen fertilizer for your garden, free as the air! Or at least as cheap as a pack of fava seed.

Some people grow favas as a "green manure" and chop up the stems and roots and mix them into topsoil. But the green tops can be steamed and served like spinach; the pods can be cooked at their young, slim stage like green beans, or left to swell to make tasty plump inner beans to cook up as you would limas. Fava beans can also be dried for use in winter soups, such as classic Italian minestrone. Choice varieties for cooking come from European seed sources (see Sources).

wildfires. Common soil additives include ashes from your fireplace or wood stove, manufactured ash (sulfate of potash), bone meals, and finely ground crushed granite, a potassium-rich rock.

Iron is a mineral that helps produce green chlorophyll in leaves. It naturally occurs in many soils, but can be depleted where concentration is low. Iron deficiencies are common in evergreens and may be identified by yellowing leaves. Animal by-products, such as blood meal, contain some iron, but it is most readily absorbed by garden plants in the form of chelated iron powders.

Calcium affects flower and fruit growth. Blossom end rot, the dark soggy spot that disfigures the bottoms of tomato, eggplant, and pepper fruits, is a sign of calcium deficiency. Calcium can be added in the form of crushed eggshells, ground oyster shells, bone meal, or dolomitic lime.

Magnesium acts as a hormone trigger to promote flowering in plants. An old gardener's trick that works is to sprinkle a teaspoon or two of Epsom salts (magnesium sulfate) around the roots of rosebushes, flowering shrubs, and tomato plants in springtime to prompt early and lavish bloom. Don't do this more than once a year—an overdose of magnesium is toxic to some plants.

Sulfur, copper, zinc, boron and *molybdenum* are trace minerals that build up a plant's disease resistance. They are usually available in the native soils of trees and shrubs. When these plants are grown outside of their native area, trace mineral supplements can be provided by rock dusts or mineralized chemical fertilizers.

In addition to adding fertilizers that contain specific minerals, some organic gardeners advocate adding various vitamins, enzymes, and herbal extracts. The jury's still out on the usefulness of adding these non-mineral soil supplements, although the organic gardening movement has prompted agricultural researchers to begin testing.

Studies by the University of California, for example, could find no benefit in the application of vitamin B_1, which is widely sold to gardeners as a growth stimulant. If you care to look at the label, what you will find is that most commercial plant supplements sold as B_1 also include rooting hormones, which do stimulate plant growth. (See Chapter 6.)

Reading a
Fertilizer Label

Fertilizers sold in the U.S. are required by law to be labeled with a chemical analysis that says what sorts of minerals and nutrients are present, and in what amounts. Labeling laws vary from state to state, but the most important information are the three big numbers (0-10-10, 5-10-5, 16-16-16, etc.) that indicate the relative amounts of available N-P-K (nitrogen, phosphorus, and potassium). Labels may also indicate the molecular form of nutrients and trace minerals if they are included (potassium phosphate or iron sulfate). Specialty fertilizers for acid-loving plants may include pH adjusters, such as sulfur.

A rather shocking realization is that most bagged bulk fertilizers are practically all filler. A bag of processed chicken manure labeled 5-2-1 is only five percent usable nitrogen, only 2 percent usable phosphorus, and one percent usable potassium; the other 92 percent of the bag is simply filler, sometimes labeled "inert materials." The filler may be barnyard debris, useful as a soil fluffer-upper, or it could just be rocks and stones. Usually, it's a little of both.

Small wonder many people prefer smaller and more economical packages of water-soluble fertilizer concentrates sold in powder, granular, or liquid form. Popular products, such as Miracle-Gro and organic fish emulsion, contain balanced mixes of nutrients and a wide variety of trace minerals. But since these chemicals and minerals are concentrated, their overuse can cause mineral-salt buildup, which appears as a crusty white rime on the surface of soil and on the sides of containers and pots. Salts that naturally occur in alkaline soils can be neutralized by adding gypsum (calcium sulfate), which inhibits the ion exchange of sodium molecules.

I've been told that adding rusty nails or iron filings to the soil is good for plants, especially tomatoes. Is this true?

Eventually, yes, when oxidation and the slow work of soil bacteria break down the metal into usable iron molecules with a positive ion charge. But the process can take years. (In the meantime, you might consider getting a tetanus booster shot.) A more effective method of getting iron to plants is to apply chelated iron, which will be absorbed by plant roots in just a few days. Applied as a foliar spray, iron supplements are also absorbed by plant leaves directly.

Should I be worried about lead in my soil?

Yes, if you garden within 75 feet of a busy street or your house was built before 1940 and may have been painted in the past with lead-based paint. Lead from paint chips and auto emissions persists in the soil and is rarely leached away by rains or irrigation. It is taken up by food plants, and when ingested can cause nerve damage, especially in children. Children can also be affected by playing in soils that contain lead—the poison may be absorbed through their hands.

If you suspect lead or other pollutants, have your soil tested. Raised beds with clean soil may be your best option if you want to grow food plants.

Playing with Dirt and Planting in Pots

One delightful aspect of gardening in containers is that you can manipulate soils to your heart's content. With a wide mix of gravel sizes, you can imitate the scrabbly scree of a glacial morain, the better to grow choice alpines or ground-hugging heathers. You can grow the miniature fruiting cranberry in a box decked out like a soggy bog, acidified with sulfur and pine needles. You can grow seven rosebushes in seven different plastic tubs, fertilize each one with something different, then see which fertilizer seems to work the best. It's easy to do a little experimenting when you're working with a pot of soil, rather than a whole garden.

Gardening in pots is also great for gardeners who have no ground, and must make do on rooftops, windowsills, balconies, porches, and decks. Adventurous gardeners who want to push the envelope will always grow some plants in pots, especially tender tropicals, such as citrus or orchids, that need to be overwintered indoors, safe from freezing weather.

Use the lightest and fluffiest soil mix you can, incorporating as much organic matter as possible (coffee grounds, compost, rice hulls) to keep it fluffed. I find many bagged container soil mixes

Some ladies asked me why their plant had died. They had got it from the very best place, and were sure they had done their very best for it....They had made a nice hole with their new trowel, and for its sole benefit they had bought a tin of Concentrated Fertilizer. This they had emptied into the hole, put in the plant, and covered it up and given it lots of water and—it had died! And these were the best and kindest of women, who would never dream of feeding a newborn infant on beefsteaks and raw brandy. But they learned their lesson well, and at once saw the sense...in rich and well-prepared garden ground such as theirs strong artificial manure was in any case superfluous.

—Gertrude Jekyll, *Wood and Garden*

are too fine for use in outdoor containers, and tend to compact down, despite the presence of lighteners, such as perlite or vermiculite (heat-expanded rock particles, usually mica). Sand, gravels, and garden soil are often too heavy in weight to be used if you plan to move the container around yourself, or will be placing it on a deck or balcony.

A good compromise is to fill the bottom third of your container with a filler of lightweight polystyrene packing pellets or recycled, ripped-up bits of the polystyrene packing material (i.e., Styrofoam) that came in the box with your new radio or VCR. Polystyrene is neutral to plants and to your soil mix, which can be a half-and-half mix of bagged potting soil and richly organic homemade compost.

> "I put away this disgusting dish of old fragments, and talk to you of my peas and clover."
>
> —Thomas Jefferson, Letter to George Washington, 1796

Don't sneer at using Styrofoam; it's the choice of many professionals. In downtown San Francisco, for example, you can walk the rooftop gardens of the Yerba Buena Center complex over vast lawns and flower gardens landscaped just a few inches over Styrofoam blocks. Two feet thick, the blocks support and spread the weight of people strolling, yet at the same time provide little extra weight on the roof of the meeting rooms of the underlying conference center.

Container gardens should be watered with care (see Chapter 2) and fertilized lightly, so that accumulated mineral salts don't burn plant roots. Fast-growing plants prefer deep containers. Shrubs and trees in large tubs should be decanted out every three to five years, to prevent their roots from circling round and round into a fibrous mess. Repotting should be done during a dormant season, and roots may be disentangled with a chopstick or garden fork and then trimmed back with sharp shears. If you trim back the roots by more than a third, you should also prune the top branches so the tree will remain well-balanced. (For more techniques on outdoor potted plants, I recommend Linda Yang's book, *The City and Town Gardener*—see Sources.)

Dishing the Dirt

I've ended this chapter talking about plants in pots because containers are to me the most fun way to play games with soil.

For years, I've grown copious quantities of blueberries in terracotta pots because my garden soil is too alkaline to support such acid-loving bushes. My blueberry pots get acidified with a yearly dose of aluminum sulfate, a harsh additive I prefer not to put into the generally organic soil of my backyard. But what I really love is being able to reconstruct the natural soil conditions of a particular species, for their benefit and mine.

On my back deck, I grow beach plum (*Prunus maritima*) in five-gallon tubs simply because it reminds me of the eastern seaboard where it grows wild along the beach. My beach plums are shaped as a pair of twin topiary standards and a delight in every season: balls of white blossoms in late spring, followed by tiny, tart purple fruits and later, red fall leaf color. Their soil mix is nearly all sand. As a bow to their likely native source of soil nutrients, the trees get fertilized with liquid fish emulsion and are mulched with handfuls of seaweed—gathered on summer trips to Fire Island, rinsed, then bagged in plastic for the plane ride home—in my attempt to supply trace minerals a beach plum might prefer.

When I close my eyes it sure smells like a day at the beach. That's fine and, in fact, something that makes me happy all year round. The plum trees have been growing well and fruiting well for five years. I guess that means they are happy, too.

When I was a rooftop terrace gardener, spring always started with dragging heavy bags of pungent cow manure past disdainful doormen and coifed neighbors onto the elevator and out through the apartment window (the builder having succeeded in cutting costs on one of my terraces by omitting the door). I was simply preparing my soil for planting.
—Linda Yang, *The City and Town Gardener*

Cycles and Seasons

4 It's Not Your Attitude; It's Your Latitude

AUGUST 8 IS ALWAYS a red-letter day on my calendar. As long as I garden through the foggy summers of San Francisco, I know I will always have my first good crop of red, ripe tomatoes by that date.

How do I know this? For years I have kept a garden diary, which invariably notes that tomatoes came in the second week of August. But I also know that thousands of gardeners—within my city and approximately 25 miles into the outlying suburbs—are also enjoying their first really good tomatoes at the same time.

And how do I know that? Some years ago I was asked by a local newspaper to come up with a gardening contest, so I devised a race—a tomato race—that would give a prize to the reader who could grow n contrast to our four calendar seasons of spring, summer, fall, and winter, the desert Cahuilla observed eight seasons, each relating to the growth cycle of mesquite, an important food plant.... The Maidu distinguished the seasons of flower time, dust time, seed time, and snow time, while Coast Miwok seasons included a ground coming out season, a hot season, a short day season and a fourth that time has forgotten. Root time, fire gone time, hot day time, and leaf on top time are the four seasons reportedly observed by the Cahto.

We live a very different kind of life today.... We do, however, have our own unique seasons to which many modern people are finely attuned. They are called baseball season, football season, basketball season, hockey season, ski season, and tourist season— to name a few.

—Linda Yamane, *In Full View*

the first vine-ripened tomato of the season. Eventually I had 75 home gardeners (including a crack team of five "seeded" contestants, the newspaper's official Tomato Racing Team) growing 225 different varieties of tomatoes. They grew heirlooms, modern hybrids, the latest imported varieties from the Soviet Union, pink tomatoes, yellow tomatoes, potato-leafed types, and long viney kinds.

Our winner, Christine Smith, weighed in in early June with a fully red fruit off an 'Early Girl' she had bought as a seedling from Woolworth's. An April 23 entry was disqualified because it failed to meet the two-inch size criteria, and only a few other contestants managed to get a ripe tomato by mid-July, in one of the foggiest summers on record.

When all the data was collected (after a giant tomato-tasting at the San Francisco County Fair), it was remarkable to learn the tricks people tried in their efforts to grow the earliest tomato. What was even more remarkable was the discovery that 70-odd gardeners, using different seeds, different soils, different fertilizers, and different techniques, all got their first flush of big red ones at approximately the same time—during the second week of August.

Sometimes Mother Nature will not be rushed.

Time and Again

Within every seed, there is an embryonic plant, curled up with its cotyledons and little root hair, seemingly asleep until the day it stirs and stretches to take the first steps to becoming a lilac bush or a championship tomato. But how does it know when to wake up? Why does it take a gardener three days to germinate radish seeds, but weeks to sprout an avocado pit?

Each plant's germination schedule is locked into its genetic code, in the DNA that is a part of the seed's very cells. The genetic code also controls when the plant will grow leaves, when it will grow flowers, and if or when it will make seeds and fruits. It's like each plant species has its own little clock, and its own little calendar.

Sometimes it doesn't take much to "start the clock." The accommodating radish seed only requires a bit of water and just a little sunlight to swell and break its dormancy. Plants in climates of heavy snowfall have evolved hard seeds, which can only be broken by the physical pressure of freezing and thawing—the routine weather schedule of a mountain springtime. A seed of tree peony, native to the Himalayas, has such a tough shell it requires two or three years of this abuse before it decides it likes the neighborhood enough to crack open. The seed capsule on a Monterey pine is so tightly wrapped it won't even drop its seeds unless the capsule—and its mother tree—catches fire.

Plants have learned to wait for the optimum season to sprout and grow. While their genetic calendars have been preprogrammed, it takes outside forces to prompt their genes to react. The major factors for prompting include the amount of sunlight and the weather—routine periods of heat or cold, moisture or dryness. Minor factors include the availability of nutri-

Monterey pine

ents (see Chapter 3) and outside agents, such as bees and other insects who assist by pollinating flowers, a first step to fruit and seed production.

The cycle of the seed is often celebrated in song and literature, and many gardeners also seem to find a rhythmic poetry in the sexual nature of plant reproduction, where the female part of a flower (pistil) holds its egg-like ovule in waiting to receive the sperm-like cells of

Wisteria pods

pollen from the male part of a flower (stamen). The zygote they produce will grow into the child fruit, a fruit that literally holds the seeds of that plant's future generations.

In the more prosaic world of competitive tomato-growing, getting a tomato plant to produce flowers earlier in the year improves the odds of getting an early-developing fruit. Christine Smith's winning techniques to prompt early flowers included a dose of magnesium to trigger flowering (see Chapter 3, page 70) and choosing a variety, 'Early Girl,' that was genetically coded to fruit early in the year.

The ripening of tomatoes is a different matter, since ripening in tomatoes and most fruits is caused by a plant-produced hormone called *ethylene*. The production of ethylene is triggered most often by a change in day length. As summer wanes, hours of daylight decrease; this factor was a constant for all who participated in the Great Tomato Race, since they all lived at the same latitude, 37° north of the equator. That is why nearly all the contestants wound up with ripe tomatoes at exactly the same time.

The Importance of Daylight and Day Length

Gardeners realize that plants need a certain amount of sunshine for their life processes. Most of the vegetables and flowers that we like to grow need at least six hours of direct sunlight per day. Some plants are adapted to shady situations and can get by with less, or with indirect light.

Explaining Photosynthesis

Just how a leaf captures light is a matter of quantum and wave mechanics, and is easily explained. (Oh, yes it is—stay with me here.) In our post-Einstein world, scientists accept that light has a dual nature: it travels as a continuous wave and also as individual particles, called *photons*.

The white light of sunshine is made up of light of different colors—a color spectrum. We can see those colors when sunlight shines through raindrops to make a rainbow. You can think of the colors of the rainbow as light waves where each different color is vibrating at a different frequency. Or you can think of those colors as groups of photons, where each different color is made up of photons of a different energy. When you are talking about photosynthesis, it's easiest to talk about light as photons.

When the sun shines on a green leaf, certain photons (particularly those from the red and blue parts of the spectrum) get absorbed by the molecules of chlorophyll. When a molecule of chlorophyll absorbs a photon, the radiant energy of the photon causes the outermost electrons of the chlorophyll molecule to become "excited" and rev up to a higher energy level. The chlorophyll molecule passes its excited electrons along to another nearby molecule. That molecule, in turn, passes the excited electrons to still other molecules, in what's called an "electronic transport train."

Every time the electrons move from one molecule to another, a little of their energy is converted to chemical energy. Then that chemical energy is used to rearrange carbon dioxide and water to make simple sugars and oxygen.

The simple sugars created by photosynthesis may remain in the leaf, to be broken down again to release energy for other chemical reactions as needed. This is the plant equivalent of grabbing a candy bar when you need a little energy boost in the afternoon. Sugar molecules also get carried around the plant through its vascular system, a water-based transport network. (See Chapter 2.) The simple sugars may also bond together to form larger and more complex molecules, such as *disaccharides* (including sucrose, the sugar in your sugar bowl) and *polysaccharides* like starches and cellulose.

Plants tend to squirrel away their complex carbohydrates into certain plant parts—seeds, for example, because the seeds will need the stored sun energy for their first growth spurt. Some plants store their energy reserves, in the form of starches, in underground tubers. This is the plant equivalent of keeping a candy bar in your pocket for that late afternoon energy boost. Every time you bite into an ear of sweet corn, or spoon into a sweet potato, you are eating the stored energy of the sun.

The cells of green leaves contain a green pigment called *chlorophyll*. Chlorophyll absorbs radiant energy from the sun's rays and, through a process called *photosynthesis*, uses that energy to convert molecules of carbon dioxide and water (taken in via leaf and root activity) into sugars such as glucose. Energy in stored plant sugars is later released and used for plant growth processes. A plant can store energy in chemical form for quite a long time, even after it's dead. The ease at which such chemicals turn back into energy is readily apparent whenever you burn the logs from your old apple tree in your living room fireplace.

There is another important plant pigment that appears not only in leaves but in small amounts along stems and bark and in seeds. It's called *phytochrome*. According to Yale University botanist Arthur W. Galston, phytochrome is a "blue-green protein" that absorbs a certain part of the sun's rays from the red end of the spectrum. It is this protein, phytochrome, that acts as a plant's internal clock.

Phabulous Phytochrome

Gardeners, farmers, and botanists had long suspected that the yearly sun cycle triggered something in the seed cycle. But it was not until 1920 that basic research began to find out exactly why.

Arthur Galston's brief on phytochrome (exhaustively documented in his book, *Life Processes of Plants;* see Sources) is that this protein not only takes in photons, but it records and calculates how many hours in a day it is taking them in. How phytochrome cells can count is something quite amazing.

A general explanation is that a plant prepares itself for winter using two bits of information: the calendar it has stored in its genes since it was a tiny seed, and the "clock" of its phytochrome proteins. Eons of evolution created the gene data, which has timed the plant for seed production in tune with local weather conditions; information that decreasing hours of daylight (occurring in autumn) precedes winter weather may be stored in the gene database.

The news that hours of daylight are decreasing is interpreted by phytochrome pigments, which come in two types: a pigment that absorbs energy from the red light, and a second pigment that absorbs the energy of far-red light. (Far-red light, also known as infrared light, is just beyond the red end of the spectrum of colors that our eyes can detect.) When red light strikes a red-ready molecule (Pr), the phytochrome turns into the second pigment (Pfr), and will only then accept far-red light. When it does receive some far-red light (more available in the warm-toned sunshine of a golden afternoon that is so beloved by photographers, and in the reddish hues of a sunset) the phytochrome turns back into a red-ready pigment.

It will stay a red-ready pigment overnight. But a molecule of far-red-ready pigment is not so stable; after a certain number of hours of nighttime dark, it will convert back to red-readiness on its own. Or it may break down further and become deactivated.

So every day, the phytochrome pigments are winking on and off, red to far-red and back again, depending on how much red light, far-red light, and darkness occurs around the leaf. As hours of light and darkness change with the seasons, the sequencing of the phytochrome molecules gets so rearranged that they set off a preprogrammed DNA trigger and the plant begins to produce important new chemicals that are called *plant hormones.*

Scientists have isolated certain hormones. Most important to the gardener are *auxins* and *gibberellins*, which spur vegetative growth; ethylene, which causes fruits to ripen; and abscisic acids, which help a plant become dormant in its off-season.

When the hours of daylight begin lengthening in spring, phytochrome in branches, bark, and dormant buds stimulates the plant to produce more auxins, which "turn on" the genetic switches that prompt the buds to elongate, stretch out, and unfurl new leaves and flowers. Phytochrome present in seed cases reacts to light and stimulates the seed case to split, germinating the seed. When day length decreases, changes in the phytochrome cause the plant to stop producing auxin, slowing down growth. Instead, it may begin to produce ethylene. Or it may produce abscisic acids that will "turn on" genetic switches throughout the entire plant, causing it to set seed, thicken bark, or shed leaves when the days grow short and winter draws near.

The process of creating these hormones over the course of the year appears to be quite automatic, although it is a rather complex gathering of amino acids. The plant's genetic response to a hormone is preprogrammed in its DNA and will also be automatic —even if a plant part is no longer attached to its vascular system or the hormone is introduced by an outside force. Placing a ripe apple in a paper bag of unripe tomatoes or unripe bananas will cause the unripe fruits to become ripe, because the ethylene hormone present in the ripe apple, released as a gas, triggers the genetic response to ripen in those other fruits. (Hormones and their manipulation are discussed in more detail in Chapter 5.)

Apple

Bananas

Practical Notions

Phytochrome can be fooled—anyone who has ever seen petunias growing under full-spectrum fluorescent lights knows that plants can be tricked by the glow of artificial sunlight. But most of us garden in the great outdoors and have to take the sun's cycle into account.

It takes 24 hours for our planet to complete a full rotation on its axis, and about 365 days—the amount of time we call a year—to revolve in its orbit around the energy-giving sun. Because the earth's axis is not straight up and down, but on a 23° slant, parts of the earth get an uneven helping of sunlight as it orbits and turns.

In temperate latitudes such as the United States, day length can be as short as eight hours in the season we call winter, when the lesser amount of sunlight fails to warm the air and gives us cold. In what we call summer, day length can stretch to as much as 15 hours, and the longer exposure of the sun's energy to the earth and air gives us hot days and warm nights. The change in day length between summer and winter is most dramatic at the poles ("lands of the midnight sun"), while at the equator sunlight and darkness are almost equal, year round.

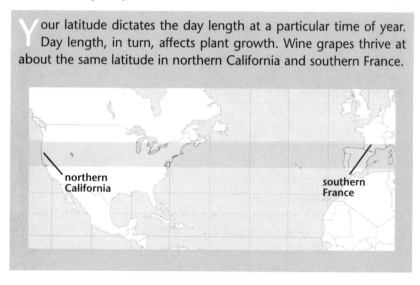

Your latitude dictates the day length at a particular time of year. Day length, in turn, affects plant growth. Wine grapes thrive at about the same latitude in northern California and southern France.

northern California

southern France

All the way around the globe, plants growing at similar latitudes have similar growing schedules. It is no coincidence that California's wine-making regions north of San Francisco produce great cabernet and chardonnay grape harvests; after all, they are at nearly the same latitude as the great wine-making regions of southern France.

Difference in day length helps to explain why it's sometimes so hard to grow certain plants in your own garden. Everybody admires

Day Length and the Dahlia

My favorite example of the risk and thrills involved in attempting to grow species out of their native latitudes is *Dahlia imperialis*, the giant tree dahlia of Mexico. Considered the world's largest herbaceous perennial, tree dahlia dies back to the ground each winter, but easily achieves segmented flowering stalks that are 30 or 40 feet high in a single season. The flowers are enormous, gorgeous, 12-inch-dinner-plate–diameter daisies of pink, purple, or white. A tree dahlia in full bloom is an amazing sight. (See Sources.)

In North America, a tree dahlia usually blooms just around Thanksgiving. This plant is so sensitive to day length that you can't make it bloom ahead of schedule. Every year, there's a race between the maturing of the flower and the first killing frost, which collapses the whole plant. Stems come crashing down, flowers and all, and that's it for the year. Since most of the U.S. has frosty days prior to Thanksgiving, tree dahlia just isn't considered practical as a garden subject (which is why you've probably never heard of it). It is grown in mild-winter regions as a novelty, and in its native Mexico it does not mind a touch of winter frost. When the segmented stems crash down and break apart, each individual segment roots and becomes a new plant by spring.

the gorgeous summer flower borders the British can produce, but their exuberance of phloxes, delphiniums, roses, lilies, and such has a lot to do with the fact that Britain is at a higher latitude. British gardens get several more hours of sunshine in the summertime than most American gardens, which is why American gardens can rarely reproduce similar flower borders. It is no coincidence that the U.S. region where English-style flower borders do flourish is the Northwest; Portland and Seattle sit above the latitude 45° north, as do London and Giverny.

Twice a year, at the spring equinox (March 21) and autumn equinox (September 22), the day length is the same all around the planet—12 hours of night and 12 hours of day. Long before phytochrome was known, before it was even, as Dr. Galston kids, "a pigment of the imagination," farmers and gardeners recognized the equinox as an important benchmark for timing crops.

Annuals, biennials, and perennials (which in this argument would include trees and shrubs) respond to day length in different ways. Annuals, such as peas, nasturtiums, and lettuces, appear to be day-neutral; short-lived opportunists, they will grow at any time of year. Shorter day lengths in wintertime do tend to slow growth down, since the hours available for photosynthesis are fewer. Genetic triggers that are more important to the flower and seed production of annuals are weather-related: a sudden burst of hot weather in spring is likely to cause peas to pod up all at once, and make lettuces "bolt," that is, send out a premature seed head on a tall, inedible stalk.

Biennials need both a day-length change and a weather change to trigger flowering and seed production. Usually spring bloomers, biennials flower and set seeds that will germinate in soil over the summertime, and when seeds are dispersed the original plant will die.

In their first autumn, the newly seeded biennials grow as ground-hugging rosettes of leaves. Such leaves are quite sturdy and will often remain green beneath winter snows.

Nasturtium

Day Length and the Gardener's Mood

Of course, day length affects animals as well as plants. The crankiness, sleepiness, and general lethargy many people experience in wintertime is, in its extreme, recognized as a malady known as *Seasonal Affective Disorder*, or SAD, which is said to be caused by wintertime's short daylight periods.

SAD is so depressing to some people that doctors recommend they buy full-spectrum light bulbs that imitate sunshine, to help restore a mental balance with additional light. In the old days, doctors often suggested their temperate-zone patients take wintertime vacations in places like Miami or Greece to get themselves cheered up. Not only did these warmer climes offer more winter sunshine, their latitudes are nearer to the equator, so the days were also longer.

After a winter vacation to a southern clime, people usually do feel peppier. That warm sun on the back of the neck stimulates a human's pineal gland, which regulates some important body hormones. Some reports on full-spectrum lights for SAD sufferers indicate that certain folks experience an increase in their sex drive. The effect may not be much different from what happens to petunias grown in the wintertime under full-spectrum fluorescent lights; the petunias will obligingly prepare for sexual reproduction by forming their flowers ahead of schedule.

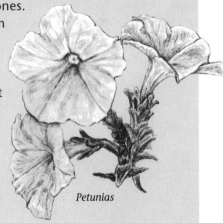

Petunias

Exposure to winter cold is a genetic trigger that sets the stage for the next phase; it is only after biennials are exposed to a period of cold temperatures that the phytochrome kicks into gear and begins sending out messages about the increasing daylight hours of spring.

This is what causes overwintered radicchios to bolt in the early spring. Other edible biennials harvested only for their roots or green vegetative parts, not the fruits or seeds that would come later, include carrots, beets, onions, cabbage, mustard, kale, and collard greens.

Cabbage

Perennials from temperate zones respond to day-length changes by initiating active growth as days grow longer, and shutting down vascular activity as days grow shorter. Most deciduous trees react to shortening days by forming a callused wall, called an *abscission*, between the leaf and its stem. The wall interrupts vascular flow and eventually the leaf falls off, leaving a sealed scar where it had been attached. Abscission is caused by abscisic acid, a hormone produced by phytochrome.

Tropical plants from around the equator, where day length doesn't change much with the seasons, use weather triggers, such as rainy seasons, to adjust their seed cycles. Indeed, the sexual reproduction of tropical species is often a leisurely process. Tropical orchids hold their flowers for months while waiting to be pollinated. Big seeds, such as the nut of the coconut palm, may take two years to develop.

Growing species native to tropical latitudes is a common practice for gardeners in temperate zones. We solve the problem of day length for some tropicals by growing them as houseplants; orchids, for example, are often grown at home or in a greenhouse under artificial light. Some tropical perennials we treat as annuals: petunias, impatiens, tomatoes, green peppers, and gladiolas.

We are so trained to think of these plants as annuals that when the shorter days of autumn arrive, we dig them up and throw them away because they've stopped growing and have gotten all raggedy looking—unless a killing frost wipes them out first. In a mild winter climate, such plants would cling (unattractively, it's true) through the darker months and launch fresh growth when days begin lengthening again.

Weather or Not

It would not make much sense for a frost-tender plant, such as corn or pumpkin, to germinate seeds immediately after it ripens, because any young plants would die in the cold of the coming winter. Corn and pumpkin protect themselves from premature sprouting by setting their clocks to "go off" only when certain conditions are met. In the case of corn, the succulent seed kernels on the cobs will shrivel into hard knobs as the cornstalk matures, hard enough to survive a cold winter. The seed plumps up again with water from spring rains and melting snows, and softens to sprout in late spring.

Pumpkin protects its seeds with a well-insulated, thick-skinned fruit that will keep in your larder (or on your front porch) for five or six months; left on the ground to rot, the fruit gradually deteriorates to release its thin-shelled seeds when warmer spring has come around again. Corn and pumpkin are both well adapted to their native climate of the American Southwest, where the weather pattern is one of freezing winters and spring floods, followed by hot, dehydrated air in summer and autumn.

With temperate-zone fruit trees, the hormone triggers are day length plus the amount of cold weather during any given winter. Apples, pears, and European plums need a longer, colder winter to set fruit from flower buds; quince and Japanese plums will produce reliably even where winters are quite mild.

See for Yourself

Because plants need so many hours of direct sunlight to grow properly, knowing where the light hits and where the deep shadows fall in your garden can help you get the most out of your growing season.

1 To give yourself a heightened awareness of shadows, try this shadow game with your kids or with your friends:

First, find a place on the ground where a sharp, straight-line shadow falls—ideally the outline of a house wall or tall fence. Using a piece of chalk, mark the line. (If the shadow falls on grass, mark it with a line of horticultural lime or peg down a length of colored ribbon.)

Guess where the shadow will be in 15 minutes. Mark your guess with a dotted line (or use a different colored ribbon).

Wait 15 minutes and see where the shadow really falls. How close was your guess? To make this a competitive game, have different people mark their guesses with different colors of chalk, or peg down different colors of ribbons, and play as many rounds as you can before the sun sets.

2 Because of the angle of the earth's axis, shadows are shorter during summer, and longer in the winter, with the longest shadows falling during the shortest day of the year, the Northern Hemisphere's winter solstice (December 21). Therefore, a building or large tree that does not shade your vegetable garden in the summer may in fact create a large, obstructive winter shadow that would mar your efforts to plant a fall crop of kale or radishes, or an early spring crop of arugula and French-style lettuces.

To determine how long winter shadows will be, use the following formula: 90° minus 23.5°, then minus your latitude in degrees. The first part of the calculation compensates for the earth's 23.5° tilt. The second part takes into account your particular location on the globe. In San Francisco at the Exploratorium, the latitude is 37°, so the calculation is (90° - 23.5°) - 37° = 29.5°. This 29.5° represents the angle of the sun, crossing from east to west at its highest point during the shortest winter day on December 21. With the sun barely reaching a third of the way into the sky at lunchtime, it's easy to see how even a medium-tall shrub, hedge, or fence sited on the south side of a property might obliterate most of the available winter sunlight to a flower bed on its northern side. The oft-repeated advice to give tender plants "a southern exposure" or "south-facing wall" attempts to maximize winter daylight hours for these species.

You can use the above calculation to find out just how long the winter shadow of a shrub, hedge, or fence might be. Just tie a long string to the top of the object and run it out in a northern direction, at the sun-angle (29.5° in this case) towards the ground. Then you will see how much shadow the object will cast on its northern side in mid-winter.

Paying attention to the angles of shadows and sunlight can pay off big when you landscape your property. For example, planting shade trees on the west or northwest side of your home will keep the house walls cooler in summer, reducing your air-conditioning costs. You may also find that ground deeply shaded by evergreens during the hot summertime receives enough low-angled sun in the winter to support winter-blooming flowers, such as Persian iris, hellebores, and crocus.

> esterday I sat out of doors near the sun-dial the whole afternoon, with the thermometer so many degrees below freezing that it will be weeks finding its way up again; but there was no wind, and beautiful sunshine, and I was well wrapped up in furs. I even had tea brought out here, to the astonishment of the menials, and sat long till after the sun had set, enjoying the frosty air. I had to drink the tea very quickly, for it showed a strong inclination to begin to freeze.
>
> —Elizabeth von Arnim, *Elizabeth and Her German Garden*

Pomologists have determined the approximate number of "chilling hours" (hours of temperature under 45° F) for popular fruits. A 'Macintosh' apple needs about 1,000 chilling hours to set fruit, while 'Anna,' a hybrid apple developed for homeowners in Florida and southern California, needs less than 200 chilling hours.

The amount of hot weather during periods of critical bud formation can determine how many flowers you will get from broadleaf evergreens, such as azaleas, camellias, and orange trees. If you grow these in containers, you can move them to the warmest part of the garden during July and August, when bud set begins, to improve next spring's floral show. Lilacs, traditionally one of the fussiest garden shrubs, demand both a warm summer and a chilly winter to bloom their best.

Learning to Read Nature's Calendar

The vast differences in weather and latitude day-length changes make it almost impossible to set anything like a universal calendar of gardening chores. Regional guides are the best; nothing should be trusted less than an old British gardening book. A book written by someone who lived in Boston or Canterbury, telling me to prune my dormant roses in March, or perhaps in November, is not going to be any help to me in northern California, since my roses are up and blooming by March and nowhere near dormant in November, unless I have helped them along by cutting off all the leaves. If you live in Detroit, it will be no help to you if I reveal that my success for getting early peas is to plant them in the ground in October; even Dr. Kevorkian would not approve of such a Michigan plant suicide.

You have to be careful with any rules of thumb. A commonplace one from the English gardener's canon is that "when you can set your foot upon twelve daisies, spring has arrived." Sometimes I have heard this quoted as "six daisies" or "three." Those crusty old English gardeners were not talking about the members of *Compositae* we find in our own spring fields; the daisy they mean is quite specifically the tiny English daisy, *Bellis perennis*, which dots grass lawns with dime-sized blossoms; it is quite easy to squish a dozen at one time even with a size-7 shoe.

More broadly adapted European weeds may be better guides: An old Italian saying is that when dandelions bloom in open fields, it is safe to plant tomatoes, which prefer a sun-warmed soil.

Native Americans who taught European settlers to plant their corn and pumpkins and tepary beans—all crops considered to be warm-season vegetables—also passed on a traditional method of determining the earliest time to sow the seed. "When the ground is warm enough to sit upon naked and comfortable," they advised, "then the earth is ready for the planting of corn."

Nothing frightened me more when I started growing roses than pruning...the first sign the bush is bouncing back is when you spot swollen eyes where new growth is to appear—at the juncture of the leaf formations and the stems on which they grow. When they are dormant, they may not be visible to the human eye. When their growth begins, they swell, turn red, and become more obvious. When you arrive with pruning shears, you must find these landmarks in order to make cuts in the right places, so make it easy on yourself and give nature a chance to help you.

—Rayford Reddell, *Growing Good Roses*

Gardening by the Moon Signs

Look through the *Old Farmer's Almanac* and you'll find charts that show not only the monthly phases of the moon, but a daily guide to the moon's position in the zodiac—along with instructions on how to use these charts to make your garden thrive.

Planting by the moon's phase is supposedly linked to the moon's gravitational forces, on the theory that if a full moon can cause high water and rising tides through the force of its gravitational pull, surely it must affect all that water coursing through the cells of every plant.

Moonlight has been shown strong enough to sprout seeds, and to

Night-blooming cereus

stimulate phytochrome in some plant species. Night-blooming cereus, which are a group of western cactuses, time their flowers to open only when the moon is full or near-full during the summer months. Other salutary lunar effects have, unfortunately, never been seriously researched.

In the days before calendars were common, and folks knew their seasons by what constellations appeared in the course of the year, I suspect a regular sky feature, such as the moon's reliable old phases, was a helpful way to jog the memory around the farm. I do check the *Old Farmer's Almanac* each month as a pattern for garden chores, mostly glad that the moon's phases allow me to procrastinate the tedious pulling of weeds until the last quarter—when they are supposed to be easier to dislodge during the blackest phase of the lunar cycle. If nothing else, the *Old Farmer's Almanac*'s moon signs provide a monthly "to do" list, and any schedule that has you pulling weeds regularly once a month is probably a good schedule to keep.

January	Full Wolf Moon
February	Full Snow Moon
March	Full Sap Moon
April	Full Sprouting Grass Moon
May	Full Flower Moon
June	Full Strawberry Moon
July	Full Thunder Moon
August	Full Sturgeon Moon
September	Full Harvest Moon
October	Full Hunter's Moon
November	Full Beaver Moon
December	Full Long Nights Moon

—*Old Farmer's Almanac*

Good plant science depends very much on observation. Phytochromes in all the plants around you, particularly native species, can signal the turn in the gardening year. A good indicator plant for North American latitudes is the hardy black locust, *Robinia pseudoacacia.* A fast-growing species grown for firewood and its canopy of fragrant white spring flowers, the black locust is usually the very last tree to leaf out in springtime, which makes it a very good indicator of favorable weather and day-length conditions that more or less guarantee that spring has arrived for good.

Black locust is also the first of the trees to begin shedding its leaves in the autumn, sometimes as early as the first week in September. Therefore, it becomes a signal plant for day-length changes, indicating that autumn has officially begun.

Fooling Mother Nature

5 Manipulating Cycles and Seasons

WHEN I WAS A LITTLE girl, spring in New Jersey began not with crocuses and daffodils, but with pussy willows. The first flower of the season, appropriately gray as February skies, arrived not with fanfare but a whisper as soft as the fluffy, silvery buds that seemed to suddenly appear in gardens and corner lots, down by the creek, massed as cut stems in a big glass vase on a shop counter, at the dentist's office, in the library.

Cut a branch of budding pussy willow (*Salix discolor*), put it in water, and its desire to be first out of the box in the spring-growth sweepstakes will make it sprout tiny roots in just a day or two. A sheaf of pussy willow branches brought home from a florist can form a mass of roots in water, so many that the cut stems may easily be transplanted, and you can have pussy willow in the garden if you want it.

When I moved to California, old-time gardeners I met showed me that cut branches of other hard-to-get, spring-blooming shrubs, such as daphnes, weigelias, and old, named lilacs, could be rooted in a vase that also held the stems of willow. Corkscrew willow (*Salix Matsudana* 'Tortuosa') was the favorite, because if you didn't have it at home you could always find it at a florist shop: the twisted, golden branches remain a big favorite with flower arrangers.

In a vase of water, cut willow branches can stimulate root growth in another plant cutting because their budding roots exude *auxin*, the hormone that initiates spring growth, and this hormone happens to be water soluble. Tiny traces of auxin from the willow stems get taken up by the vascular tissue of the other cutting as it draws water. Just as a plant can't tell whether it's being fed with a synthetic or organic fertilizer, the second cutting doesn't care where the auxin came from—it just takes the chemical in and starts to root.

Once botanists began to isolate auxin and the hormones responsible for plant growth, it was only natural that they would begin to explore how those chemicals might be used in new ways to grow more, bigger, and better plants. Auxin has turned out to be a gold mine for everybody, because it can be readily produced in a laboratory. One of the natural sources it can be derived from is human urine.

An oak tree, over a life span of hundreds of years, sheds tons of acorns onto the ground beneath it. From this mass, one, on average, will find itself in a place where it can germinate, grow to maturity, and in turn produce the successive crops needed to produce one more tree; two generations, perhaps 400 or more years gone by, and just two trees to show for it.
—Peter Thompson, *Creative Propagation*

The Tale of the Foolish Seedlings

I n the late 19th century, Japanese rice farmers noted extraordinarily tall seedlings rising at intervals in fields of otherwise uniform plants. Hoping these tall plants might constitute a strain of giant rice, the farmers watched the seedlings in the expectation that they would flower and produce grain to be used as seed. The seedlings never reached sexual maturity, however; instead, the stems grew unusually fast, then died before flowering. The Japanese farmers named this phenomenon the "foolish seedling," or Bakanae disease.

In the 1920s, a Japanese botanist named Eiichi Kurosawa discovered that these foolish seedlings were all infected by a fungus, *Gibberella fujikuroi*. If the spores of the fungus were transferred to an uninfected plant, the latter became diseased or hyperelongated. If the spores were placed in an artificial medium...the liquid in the culture also contained the active principle producing hyperelongation. From this liquid, Japanese scientists were able to isolate and identify the active material, which they named *gibberellic acid*.

Since then, more than 80 similar molecules, collectively called gibberellins, have been found in higher plants. Structurally, the gibberellin molecules are roughly analogous to the steroid group of animal hormones.

Rice seedling

—Arthur W. Galston, *Life Processes of Plants*

Commercially-made auxin hormone is available on your garden shop shelf in the form of rooting products, such as Dip-It or RooTone, that you can use to propagate new plants by inducing cuttings to form roots. Although in nature, several chemical substances (including the plant-produced vitamins B_1, B_6, and niacin) work together to stimulate the growth of roots, root production can be accomplished quite quickly with merely a dose of auxin—delivered to the point of a cutting where you want the roots to grow.

You can manipulate hormones in your garden plants in a completely natural way. No, I'm not suggesting you pee on your hardwood cuttings to get them to root. But you might want to get yourself some willow branches, or start working with the other hormones that already exist in your garden plants, to help them to grow more beautiful and more fruitful.

Hormones for the Gardener

Manipulating cycles and seasons to get plants to grow better and faster is really a matter of manipulating these naturally occurring plant hormones. Of all the hormones produced in response to changes in day length (see Chapter 4), auxin is the hormone most responsible for active growth. Phytochromes stimulate auxin production only in certain parts of a plant. These are called *growing points:* near the tips of the roots, the tips of new shoots, the very top or apex of a landscape tree. In large, woody trees, auxin is also created in the cambial cells of tree trunks, where it may be transported in minute quantities through the tree's vascular network.

Plants grow up, towards the light, but auxin always moves from the top down. Produced in the growing tips, terminal buds, and side buds along a tree trunk, the hormone travels towards the base of the plant, never in the opposite direction. This is because of the particular way the hormone molecule passes through plant cell membranes. Once it passes through the membrane on the down-facing (rootwards) part of the cell, it can't get back up again.

If a source of auxin is removed during the growing season, the plant's behavior will change. Say you are cutting flowers from a rosebush or pruning back a pine tree; what you are really doing is removing the auxin-rich terminal bud on those shoots. Ordinarily, other buds lower or to the side of the growing point that was cut off—buds that may have appeared weak or dormant—will begin to sprout vigorously once the terminal bud is gone.

But scientists have found that if auxin is artificially applied to the cut surface, the lower buds will remain dormant. So auxin not only creates growth, it acts as a growth regulator, making sure that stems grow in such a way as to create a balanced specimen reaching towards the light.

Tip pruning

When you cut the terminal bud and its growing point, you remove the source of the auxin that was mobilizing stems and suppressing dormant buds. Those dormant buds wake up and the vascular flow of nutrients begins, allowing these reserve buds to elongate into leafy stems that will once again reach towards the light, or perhaps create a flower.

You've probably noticed that plants seem to bend toward light. The bending effect, called *phototropism*, is actually caused by an elongation of the plant cells on the darker side of the plant. The amount of sunlight triggers phytochrome activity near the growing point, and sends along the chemical message to produce auxins, which prompt an elongation of stem cells and stretch the stem in a certain direction.

Pinching and *tip pruning* are two techniques that manipulate auxins in the growing points with a specific goal in mind—to create bushier plants with more plant tips, usually to produce more flowers or create a richly dense mass of foliage in a shrub. But these techniques only work if you cut the plant back early in the growing season, when days are lengthening, and phytochrome is stimulating the production of enough fresh auxin to launch growing points in the reserve buds. If you pinch back in autumn, when days are growing short, new auxins will not be produced, and the side buds will fail to develop. This is particularly important to know when trimming evergreen hedges: If you shear bushes back too late in the summer, you'll spend the rest of the year looking at a nicely shaped bunch of dry woody sticks, with no new growth.

To get larger flowers, gardeners use a technique called *disbudding*. This is just what it implies: if a leafy shoot displays three or more flower buds, the largest, or terminal bud, will be retained, but any smaller side buds will be removed. In roses, dahlias, carnations, marigolds, and other flowers grown to exhibition size, disbudding is common, done by pinching off the minor buds with fingernails or short clippers, well before the buds have any chance to open and bloom. If the lesser buds are not removed, the auxins in the terminal flower bud will inhibit these side buds throughout the growth period, and your result will be one good-sized (but not award-winning) blossom that has some wimpy, weak blossoms alongside.

Disbudding

Removing side buds does not cause new growth below where you have disbudded, because there is sufficient auxin flowing from the terminal bud to suppress growth quite far down the stem. And so, with the first set of side buds completely gone and lower dormant buds well suppressed, all the vascular action in that particular leafy stem heads for the terminal bud. Flush with extra nutrients, the remaining bud can grow into a significantly larger flower—large enough, perhaps, to win a blue ribbon at a

D isbud for the bloom formation that suits you and the varieties you grow. If blooms are prettiest one to a stem, pinch out the side buds as soon as you see them developing and direct all energy to the terminal [main] bud. If a bush likes to bloom in sprays, and you too like masses of bloom, pinch out the terminal bud, which otherwise will open before everything else.

—Rayford Reddell, *Growing Good Roses*

county fair. Most flowering plants that produce multiple buds react to disbudding this way.

Let's pause a minute and consider this reaction from the point of view of a plant that wants to continue its seed cycle. It's almost as if the plant says, "Well, I've only got one shot at pollination, so I better make the flower as big as I can, so the pollinating insects don't miss it." Shrewd thinking—if a plant could think. But what's really going on is that a plant's growth processes are quite elastic, and readily adapt to building more plant cells wherever they can, as long as the nutritional building blocks are in abundance.

Disbudding works in the vegetable patch, too. A gardener can remove unpollinated flowers from a tomato plant, or remove the smaller tomatoes in a stem cluster. By judiciously redirecting auxins and the vascular flow of nutrients, you can wind up with fewer tomatoes per stem, but those that do grow can get really, really big.

(The largest tomato on record weighed in at 7 pounds, 12 ounces in 1994.) Professional pumpkin growers, who compete in nationwide contests, travel around with pumpkins that weigh 700 to 800 pounds, and nearly all their prize-winning pumpkins are grown as the single fruit of a very well-pampered vine.

Orchardists also recommend removing a quarter or third of the growing apples on an apple tree in order to get a smaller crop of larger and more tasty apples. Thinning fruit is also done on oranges, pears, grape clusters, and all garden fruit trees in the plum family. The best time to do this is in the early summer, before the fruits have reached an inch in size, to direct vascular flow early enough to do some good. If the fruits are clipped off cleanly with sharp shears, there is no harm to the tree. In fact, many fruit trees routinely do what's called "summer fruit drop" and let go of their excess fruit. Beginning gardeners sometimes find this alarming.

Have you ever seen really humongous strawberries in the grocery store? These are the first fruit from a strawberry plant's central bud. Farmers call them the "King Berry" and will often flood their strawberry patches with water and fertilizers to get them to grow to the biggest size. (Alas, that excess water in the tissues surrounding the strawberry seeds often makes the King Berry tasteless, not sweet.)

Once the King Berry is picked, a strawberry plant will make a subsequent round of new buds with smaller berries. If these are picked, a strawberry will usually continue to put out new flowers and new fruits in an attempt to set and spread its seeds.

Totipotency, the Soul of Propagation

Once you realize that the stuff that makes a plant actually "grow" is only located in certain parts of the plant, many of the techniques used to divide and multiply plants suddenly make sense. Here's just one example: If you cut a length of grapevine during its dormant state, and stick it in the ground or in a pot of garden soil, odds are very good your cutting will form roots by late spring, and become a brand new grapevine plant.

This works for grapes and many other plants that can be propagated from rootable cuttings because minute quantities of auxin and a few little blobs of phytochrome pigment exist in the dormant bud and its surrounding bark. Once auxins are activated in the dormant grape cuttings (as days grow longer in spring) they flow down towards the ground and start the business of growing new roots. Dusting a cutting with a rooting hormone powder makes the job all the faster, since the auxin in the powder starts immediately making cells for new roots at the root level. Trees that root easily from branch cuttings include maple, boxwood, fig, laurel, cherry, plum, willow, and tamarisk.

Botanists call the ability of a plant to regenerate missing parts *totipotency*. As we learned in Chapter 1, all a plant's genetic qualities exist in a single plant cell. But not all plant cells have the

See for Yourself

1 You don't need a test tube to multiply many of your favorite border flowers, just a spading fork and sharp shears. Choose a perennial you would like to have more of; good candidates include aster, coreopsis, lavender, lamb's ears, Shasta daisy, monarda, and yarrow.

In spring or autumn (well before or shortly after the plant has bloomed is the best time), trim back the stems to 6 inches or so, then pry up your perennial with the spading fork. Brush off some dirt from the roots, then just tear the root section apart with your hands, or cut the root mass into sections with your shears.

Plant the rooted sections into new locations, and water as well as you would any transplant. As long as there are enough root-tips and dormant buds intact, your "divisions" will take hold and grow into fine specimens, each one genetically identical to the original plant.

Iris rhizomes, dahlia tubers, and daylilies can also be multiplied through division, but only the underground sections that are attached to a fan of leaves or show a shoot or "eye" will root, because these are where the auxin-producing growing points are. Rhizomes or tubers without leaves or eyes are old storage units and will not flower.

2 African violets are one of the few plants with meristem tissue in their leaves. Legions of little old ladies multiply their favorite *Saintpaulias* this way, and you can, too.

Cut a big leaf from a healthy African violet plant, along with a length of the leaf stalk. Dust or dip the cutting with a rooting hormone

African violets

product, and stick it into a flowerpot filled with soil. Use a pre-bagged soil mix for African violets if you wish, but above all make sure the bottom of the leaf (where the growing point is) is touching the soil mix.

Water the pot thoroughly and slip the whole thing into a clear plastic food bag. Secure it with a twist tie. Place your mini-greenhouse in a warm room with bright but indirect sunlight. After a few days, if it looks too muggy inside the little green-house, open the twist tie and let the plant air out for an hour or two. Do not let the soil dry out.

With success, several small rosettes of tiny leaves—new plants—will form at the base of the leaf. When the leaves are the size of your fingernail, remove the plastic cover. Fertilize, water regularly, and give bright indirect light until the rosettes are large enough to be removed and repotted as individuals.

ability to regenerate and create an entirely new plant from a short piece. A plant can't regenerate from cells that make up its flowers; normally a plant can't regenerate from the cells in a bit of leaf or those found in tuber storage units. (The two big exceptions here are African violets and potatoes.)

When you're looking to make a rootable cutting of a tree, vine, or shrub, choose a stem section that has a lot of bumpy little dormant buds along the sides—the cells that manage totipotency are most likely assembled in clusters near these growing points. Another name for these clusters of cells is *meristem tissue*. While scientists were fooling around with auxins and other hormones, they discovered that bits of meristem tissue could re-create an entire plant, if it could be supported with nutrients in a controlled environment, such as a laboratory flask. A new plant grown from a tiny piece of meristem tissue is genetically identical to the mother plant—it is a clone. Orchids, which are slow to propagate by other means, are routinely grown from meristem clones in orchid nurseries.

Flowering and Fruit Production

Scientists have been searching for a suspected hormone that is responsible for the creation of flowers, but they have yet to find one. Both day length and weather appear to be genetic triggers to flower production: By altering temperatures and light conditions in a greenhouse, commercial cut-flower growers are easily able to coax roses, gladiola, tulips, and carnations to bloom year-round. (That's one reason these plants are the usual staples of florist arrangements.) But flowering appears to be a result of several different chemical processes, and two hormones, gibberellin and ethylene, seem to have an effect on flowering in plants.

Biologists do know that once a flower is pollinated, the seed that is created begins to diffuse auxin into the walls of the surrounding ovary—that little swollen sac at the base of most flowers. The suppressing qualities of the auxin attempt to redirect the vascular flow of nutrients toward the developing seeds, and cells in the ovary begin dividing, creating a larger swelling that becomes a green fruit.

As soon as the plant begins producing the later-season hormone ethylene, the fruit ripens. As in the case of a strawberry, tomato, apple, or rose hip (the bright seed pod of the rosebush), the fruit becomes attractively red, soft, and fragrant. It will have a certain appeal to humans and other animals, who will eat the fruit and distribute the seeds at a distance from the parent plant.

Rose hips

To get fruit to ripen faster, agricultural scientists have experimented with applications of auxins, but they have had the most luck with exposing unripe fruits to applications of ethylene, which is a stable gas molecule made up of two carbon atoms and four hydrogen atoms. The fruits of such labors can literally be viewed in any grocery store. Ever wonder why those "vine-ripened" tomatoes imported from Holland are always the same strange and identical red hue? Sure, they're ripened on the vine—but in greenhouses pumped full of ethylene gas, which promotes an even coloring. When bananas are shipped to the continental U.S. from Ecuador, they are picked green and treated with ethylene two days before hitting the supermarket shelves; this brief exposure is enough to prompt the fruit's skin to turn yellow and its flesh to become sweeter as the stored starch is broken down into sugars.

But when roses are fresh-cut by the dozen in Colombia to be shipped by air to the U.S., the boxes they are shipped in are shrink-wrapped in plastic, and all the air is pumped out of the plastic-coated box. This is to remove any ethylene gas that may be circulating around the roses. (The gas would start ripening the roses into fruits—rose hips—and the petals would all fall off.) The boxes are then pumped full of inert nitrogen gas (the same gas that plumps up a plastic bag of potato chips in your grocery store). This process helps preserve the roses until they reach your local florist.

Rose

It is also a common practice to store apples over winter in warehouses filled with high levels of carbon dioxide, which appears to inhibit the ripening function of an apple's natural ethylene. Such apples may look rosy-red and ripe, even if they are not really ripe, because they may also have been treated with a different sort of synthetic chemical hormone that just turns the skins red.

We can use what we know about natural plant hormones to make small but significant strides in our own gardens—without a lot of fussing about with gas tanks and chemicals. You already know how to use ethylene to ripen green avocados, green bananas, and green tomatoes: Just put an ethylene-exuding ripe apple in a paper bag with the unripe fruit, place it on your kitchen counter, and wait.

This year, I plan to try the apple trick outside, to get red, vine-ripe tomatoes earlier in the season. An interesting bit of data collected from The Great Tomato Race was that almost everyone's first red tomato ripened at the bottom at the vine, close to the soil level.

Was it because that tomato was first in line for nutrients at the trough of vascular flow? Was it because the ground was warm and radiating heat to help the ripening along? Or was it because the air around the base of the plant was calm and still? Did the first gaseous molecules of ethylene hang around for a while, doing their ripening chores, instead of blowing away on a spring breeze?

My experiment will be to tie up a Granny Smith apple near the first good-sized green tomato that comes up. Will the ethylene from the apple make a difference?

Try it and see for yourself.

Deadheading and Dormancy

To keep your garden flowers blooming over a longer period, simply remove fading blossoms before they have begun to fully develop seeds. What you are doing is removing the auxin source from the developing seeds, removing that hormone suppressant that is holding dormant or lower buds in check. Once top buds are cut, the plant will try again to produce new flowers with its standby buds. "Deadheading" works especially well with annual flowers such as cosmos, which will happily keep producing new blooms from May to frost if the old ones are always cut off.

In the vegetable garden, you can practice deadheading on veg-

etables such as broccoli. The parts of the broccoli plant that we eat are the immature green flowers that grow out of a central stalk that is surrounded by a rosette of cabbage-like leaves. If you cut the stalk before the flowers open and turn yellow, reserve side shoots will grow their own secondary stalks. These stalks and their flower heads might not be as large as the first stalk, but in four to six weeks will develop enough to provide a second crop for the table.

Perennial garden flowers such as iris, which spend much of their time and energy developing rootlike rhizomes that creep along the ground to create new plants in non-seed fashion, rarely make a second try if you cut the first flowers for your indoor bouquets. But if the leggy flower spikes of biennials, such as foxglove and Canterbury bells, are cut off a few inches from the ground at the end of their spring bloom period, these plants will often send up a second round of shorter spikes in an attempt to set seed in time for their autumn germination period.

Timing is important. If you wait too long and the plant has time to produce ripe fruits and ripe seeds, the back buds will have been too suppressed to grow well. When days begin to grow shorter, the day length signals of the phytochrome (see Chapter 4) cause the plant to cease producing auxins, and growth slows. Annuals and biennials will usually die once their seeds are ripened.

Perennials that are spring bloomers usually enter a period of vegetative growth once they are done flowering. Some strawberry varieties will quit making new flowers and begin sending out runners—arching stems with growth points on the ends. Each growth point forms a crownlike leafy bit called a *stolon*, which will root and make new plants. As mentioned, the iris develops new rhizomes; wisteria and the white winter clematis (*Clematis Armandii*) send forth vigorous new leafy vines once they've stopped flowering, all the better to grow the chlorophylls and make the sugars that will sustain them during next spring's early show. Like the lilacs, rhododendrons, and azaleas, they will produce new flower buds in August and September. In the gardener's language, they are said to bloom on "old wood," because the flowers appear on stems that grew the summer before.

I think the reason that Grandma's garden is so well remembered is that children were allowed to pick as many flowers as they needed, for there was nothing rare or choice in the flower beds: petunias, four o'clocks, roses. The more they were picked the better they bloomed.

—Gertrude Lawrence, *Through the Garden Gate*

Auxins and growth points are everywhere on such plants during this flush of post-flowering leaf-making. If you would have two or four stems next year, where only one is now, cut spring-bloomers back directly after flowering, and this will encourage more leafy growth, and then more flowers next year.

Phytochrome Cues and Winter Pruning

Perennials that bloom on new wood and fruit in late summer and fall react differently. Once their fruits have ripened, growth stops and dormancy begins, thanks to the actions of another hormone called *abscisic acid.*

Phytochromes stimulate the production of abscisic acid in response to shorter daylight hours and, like auxin, the acid will travel to specific spots in a plant. In deciduous trees and shrubs, abscisic acid heads first into the *petioles,* or leaf stems, to build a thin wall of cells (called an *abscission*) between leaf and branch, which cuts off vascular flow. Eventually the leaf drops off and a scar forms, sealing the branch. Unable to make chlorophyll or pump water through leaf transpiration, the leafless specimen goes into a winter sleep.

Abscisic acid also travels into the forming seeds, where it apparently provides a protective function by not allowing germination over the winter period. (Temperature and moisture changes, such as those caused by spring thaws, break down

Hydrangeas that are never pruned become enormous and lopsided and their owners may, in desperation, cut entire bushes right to the ground. Done in spring, this really isn't at all stupid. A year's flowering will be lost, but the bush will have been rejuvenated at a stroke, and will be set, if new shoots survive the winter, to give the display of its life in the following year.
—Christopher Lloyd, *The Adventurous Gardener*

abscisic acid levels so that germination can begin at the appropriate time.)

Scientists at the University of Wales discovered that abscisic acid, when isolated and applied artificially, has the power to slow down stem growth and thicken bark on woody twigs. It is also responsible for prompting the buildup of a special, waxy coating that forms on the winter-dormant buds of deciduous trees to suppress the tender growing points within the buds. This is a protective measure that prevents buds from sprouting during freak warm spells in the wintertime. But after spring weather wears down the abscisic acid coating on the overwintering buds, they sprout to vigorous life.

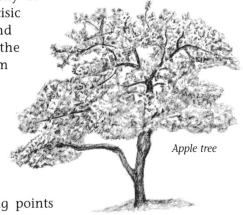

Apple tree

To increase or shape branches on trees, shrubs, and vines that bloom and fruit on new wood—our summer and fall specials, such as roses, apples, and grapes—prune them during their dormant period. New wood is what will sprout from the almost-invisible reserve buds that had been suppressed by a terminal bud that you pruned off while the plant was sleeping. Cutting back terminal buds selectively, prior to the spring growth surge, will produce healthy branching in several directions.

More Phooling Around with Phytochromes

If the art of good pruning is simply learning where the growing points are on a particular plant, the science of seed-sowing can be considered merely a matter of working out the details of day length and weather—and then, in some cases, working around those factors to get the job done.

Simple Seed Starting

C lear plastic kitchen bags and recycled, plastic-coated paper milk cartons make great seed-starting flats. Cut the cartons in half lengthwise; quart and half-gallon cartons will fit easily on a windowsill.

Fill the cartons with moistened potting soil (pre-bagged, sterilized potting mix is best) and plant your seeds according to the instructions on the seed packet.

Slip the planted cartons into clear plastic food bags, and secure them with a twist tie.

Place your mini-greenhouses on a bright windowsill. You should not have to water the flats for a while, but if the soil mix looks dry, take the carton out of the bag and water lightly.

For plants that prefer warm soil for germination (pumpkins, squashes, okra, cucumbers, basil, tomato, pepper, eggplant), place your plastic-wrapped milk cartons on top of your television set. This will provide a gentle bottom heat that will germinate the seeds more quickly.

Seed germination

When seedlings emerge, move the cartons to a sunny window. Rotate the cartons so the seedling stems don't curve. (You want them up straight, but their movement towards the light is a sign that auxins are working properly.)

When the leaves on the seedlings turn deep green, remove the twist tie, but don't remove the plastic bag. Mist or lightly water as needed to keep the soil mix moist.

When the seedlings have grown their first true leaves, remove the plastic bag. At this point you should move the cartons to a situation that is as close to natural light as possible, in order to continue growth until they are ready to be planted outside.

Starting a batch of tomato or pumpkin seeds on a windowsill in February or March is a common technique. What we're really doing is trying to approximate a normal, outdoor spring for these warm-season vegetables, but well ahead of schedule. A simple seed-starting setup (see page 112) provides all you need to fool the phytochromes present within the seed. Perhaps it really feels like spring to the seed as you shake it out of its paper packet: the temperature inside your house is warm, certainly above 65°. As you water the seeds in the flat, and keep them moist, you provide a liquid to dissolve abscisic acid on the seed-case; as the seed becomes porous, water is absorbed and the seed case swells and splits, releasing the seed.

You need light, of course. The amount of daylight needed to sprout seeds varies greatly; a few hours are all that is needed to stimulate phytochromes present in lettuce or coleus seeds, which should be sprinkled lightly on top of the soil in the flat, for they will not grow without exposure to light. Some seeds, such as nasturtium and fennel, demand periods of total darkness to sprout. But most plants, even when they have been buried in soil, will still absorb enough radiant energy from the sun to stimulate phytochromes. That's because some phytochrome pigments in seed-cases are the far-red-ready type (Pfr), only waiting for far-red light, that part of the spectrum we can't see but can feel as heat—otherwise known as infrared. Infrared radiation can penetrate some inches of soil to germinate buried seeds.

A tomato seedling will sprout if given warmth and water, but it won't come to much unless it gets adequate light. Fortunately, in its early stages almost any sort of light will do, and not much of it is required. A kitchen windowsill is usually just fine.

Once the seedling is up, day length and the color of light become a factor. For vigorous growth, phytochromes in the leaves of the plant need to experience the variable light of increasing day length.

Artificial electric light, a reality for the home gardener only for a scant 100 years, solves the day length problem beautifully. With full-spectrum lights and a timer, you can grow orchids in your basement, and could probably even grow sweet corn if you gave it a good college try.

USDA
Climate Zones

I n gardening, rules of thumb are simply that—not strict absolutes. So it is always sad to see people reject promising new trees and shrubs for their landscapes because they've been told, "Oh, that won't grow here." Or they think the USDA Climate Zone Map arrived etched on a set of stone tablets, and wasn't the product of computer calculations by the U.S. Department of Agriculture.

The USDA map was designed as a general reference for plant hardiness: It shows, generally, what the coldest winter temperatures are in different regions of the 50 states. Southern Florida, for example, resides in Zone 10, with average winter temperatures no lower than 30°. Parts of Minnesota are parked in frigid Zone 2, where -50° winter weather is not really rare.

You can find the USDA map in many how-to books and in seed and plant catalogs, which obligingly print, for each species offered, the zones where that species is known to grow well. The job of matching individual species with zones was given to the Arnold Arboretum of Harvard University.

It's no joke to say that Harvard was, well, conservative, in its judgments of cold-hardiness for many plants. Katherine Pyle, one of the founders of the amateur enthusiast's group California Rare Fruit Growers, encourages the adventurous to "stretch the envelope" at least one zone in either direction when considering a plant, and she'll often experiment with species two zones off.

Another great experimenter is *Denver Post* gardening columnist Lauren Springer. "If the USDA map were an accurate predictor of plant survival," she once wrote, "a good third of the plants I grow would be dead by now."

Be bold, dear friends, and think in terms of microclimates. Use what you know about weather, water, shadow, and sunlight patterns in your own backyard to get a jump on spring. A patch of ground just two feet square that gets a good blast of warm sun in the wintertime is a good place to try to grow the hardy winter greens that are served in trendy restaurants: arugula, mache or corn salad, cress, and mizuna. A sunny wall of your house that may also enjoy radiating heat from inside the home can be a good spot to try the fragrant flowering winter jasmine.

Where winters are warm, the shady north side of your house might be the best spot to site species that need many chilling hours, such as lilac or peony, or that 'Macintosh' apple tree you've always wanted. More portable species that need prechilling, such as tulip and lily bulbs, can get their cold treatment in your refrigerator.

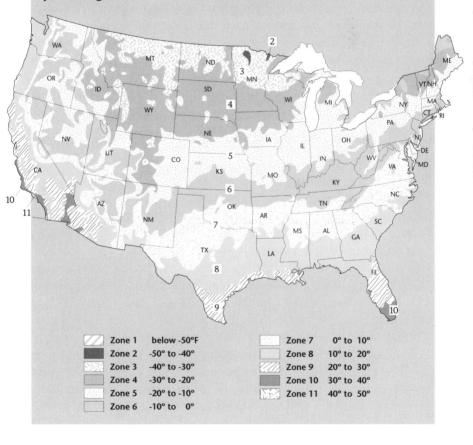

	Zone 1	below -50°F		Zone 7	0° to 10°
	Zone 2	-50° to -40°		Zone 8	10° to 20°
	Zone 3	-40° to -30°		Zone 9	20° to 30°
	Zone 4	-30° to -20°		Zone 10	30° to 40°
	Zone 5	-20° to -10°		Zone 11	40° to 50°
	Zone 6	-10° to 0°			

Outwitting the Weather

Even before gardeners could fool phytochrome with artificial lighting, they still had many tips and tricks for growing fruits, vegetables, and flowers out of season. Old gardening books are full of such tips—and mostly, they still work.

One way to perk plants up and rev up their phytochromes to the naturally increasing daylight hours of spring is to sprout them first and then grow them outdoors as early as possible in the year. The only stumbling block is cold weather—some tender species die when exposed to freezing temperatures, and not-so-hardy species do not grow well in chilly soil.

Ancient Romans solved the problem of keeping tender species warm over winter by sprouting them indoors, then planting them outside in pits surrounded by heaping piles of fresh manure. The steam and heat given off by decomposing dung was often enough to keep the local air from freezing, and prized plants could get a good head start.

Later, the builders of medieval cloisters and the designers of the great French chateaus learned to set close walls of stone and brick facing south, and in these sun-warmed, frost-protected spaces grew figs, jasmine, and roses. Today many British gardeners plant frost-tender species, such as the brilliant

> "If we garden with nature, rather than against her, our life will be easier and the result more restful in every way."
>
> —Graham Stuart Thomas, Perennial Garden Plants

yellow-flowered California flannel bush, bird-of-paradise, or Mexican orange (*Choisya ternata*), against a protected, south-facing wall to get the maximum hours of winter sun and warmth. A bad winter with killing frosts will kill off such plants, but what would you risk to be the envy of your gardening friends?

The availability of glass for housing—an innovation of but a few hundred years—makes it possible for us to enjoy an early start in the comfort of a sunny kitchen window. We can force spring bulbs and start seedlings of vegetables that we will transplant once the ground has warmed up.

A glass greenhouse or one made of heavy clear plastic provides some protection for new seedlings in the early spring, although in sub-zero climates a greenhouse heater may have to be added. Where there is not room for a greenhouse, a glass cold frame with a heating cable works. To protect individual plants, try a glass bell cloche—or its modern equivalent, the WalloWater. This is a mini-greenhouse "tent" made of plastic tubes that are filled with water. During sunny winter days, sunshine heats up the water in the tubes, which keeps the plant inside the unit warm and cozy overnight.

Fancy garden catalogs, such as the one from Smith & Hawken, still sell glass bell cloches, simple domes of heavy glass that keep out the frost and are a mainstay of French market gardeners. In my own garden I use one-gallon glass cider jugs with the bottoms cut off, though I would recommend WalloWater if April snowstorms are not unheard of in your area.

In regions where winter nips quickly at the heels of harvesttime, gardeners often worry that tomatoes and peppers or grapes will not have the time to fully ripen before they are knocked down by frost. Season extenders in this case may be a heavy tarp of clear plastic set over the tomato cages, or perhaps even a wool blanket tossed over the grapevines each night.

In warm winter areas, gardeners may have a different problem: Day length is fine, but there is no cold weather to trigger genetic changes. Some plants don't flower if they don't get enough chilling hours, and that includes some favorites, such as Macintosh apples, French lilacs, and garden peonies. But help is on the way.

Re-Dialing the Genetic Codes

There is another way to re-route the DNA calendar and clock in your favorite plant species: You can try to create an entirely new plant, a hybrid, with a brand-new genetic code.

Hybridization in the modern age is quite a wonder. As I mentioned briefly in Chapter 1, folks are busy doing gene splicing, irradiating seeds, and treating seeds with chemicals, such as gibberellins or colchicine. To discuss all that's going on would take quite a bit more space than this book has room for—and it would be entirely out of date when the next DNA breakthrough came along.

Savvy gardeners learn to pick and choose among current offerings to find varieties or cultivars, whether hybrid or wild, with useful qualities. It turns out it's entirely possible to grow peonies that don't need as many chilling hours—the trick is to select those varieties that are early bloomers, such as 'Mon. Jules Elie,' genetically coded to bloom with several weeks' less chilling time than later-blooming varieties.

If you wanted a red rosebush, you wouldn't let someone try to sell you a yellow rose, would you? So why pick up any sort of tomato plant sold in a garden center six-pack when there are about 300 different cultivars available to grow, and most general mail-order seed companies list at least a dozen?

If you're looking for a tomato that will fruit early, choose one that is genetically programmed to produce flowers early, such as 'Early Girl' or 'Early Cascade.' If your soil is contaminated with bacteria that cause vascular wilts, you'll want to choose a modern hybrid tomato variety that was bred to be disease resistant. If you garden in containers, your best bet is a dwarf such as 'Patio.'

Getting a tomato that will keep its fruit well over the winter is a topic being addressed on many fronts. Near Sacramento,

Tomatoes

the Calgene company recently unveiled a long-keeping tomato that was created by removing a genetic trigger for ripening and inserting it back into the DNA chain backwards and upside down. The result is a tomato that resists rotting and will keep on a supermarket shelf for eight weeks.

Not far away from Calgene's corporate offices there is a backyard farm where tomatoes, beans, and eggplants from all over the world are grown for the Seed Savers Exchange, a nonprofit organization that collects, preserves, and disseminates plant seeds to researchers and home gardeners alike.

Suzanne Ashworth, the curator of the Seed Savers Sacramento farm, once showed me, with some excitement, an eggplant cultivar she was growing there. The seed had come from Vietnamese immigrants who had settled in Sacramento. The fruits of this eggplant were hard, green, and bitter, and were used as a condiment.

"*To be a successful gardener, one must be a realist.*"
—*Vita Sackville West*

The farm had recently experienced its first few frosts (Sacramento does get freezing weather, even though it is in USDA Zone 9) and one of the few plants left standing among the solanums—tomatoes, peppers, and eggplants—was this Vietnamese import. It had clearly weathered with aplomb the kind of freezing that typically zaps tomato and eggplant vines to their deaths, and its little green fruits stood firm.

As Ashworth explained, since tomatoes and eggplants are from the same plant family, it is entirely possible that one day genetic material from this vegetable survivor might be used to develop a tomato plant that would not die from cold. And then who would want a tomato that had sat on a supermarket shelf for eight weeks, when you could pick a fresh one from the garden even on Christmas Day?

Ecology and Habitats

6 Partnerships with the Natural World

A N ECOLOGICAL approach to gardening may not give you bigger tomatoes or sweeter sweet corn, but it certainly makes a landscape more interesting.

The second season after I completely stopped using a liquid chemical fertilizer on my backyard soil, I noticed little brown salamanders when I bent down to weed. These creatures are sensitive to high levels of soil salts, and my new choice of fertilizer (lightly composted turkey manure) was more to their liking. Since salamanders eat snails and slug eggs, my snail and slug populations dropped. Now all my garden plants can be grown without frequent applications of poison snail baits.

I seem to get more garden by doing less work. In the flower borders, for example, I leave self-sowing biennials, such as lychnis and toadflax, and native perennials, such as wild buckwheat, to go to seed each autumn. This strategy brings in cheery flocks of winter finches, along with the occasional field mouse.

I never really appreciated those mice until one winter's day when a big red hawk came to rest on the top of my back fence, just ten feet away from where I stood. If you live in the country, you probably see hawks all the time. But I live five minutes from downtown San Francisco and what's behind my back fence is a steep drop to a busy freeway.

Hawks have always flown around the hills of San Francisco; they are part of the natural landscape in my part of the world. But when so much of the human landscape seems to look the same, with a chain store or gas station on every corner, it was wonderful to be reminded by a visiting hawk that natural diversity persists—even in the most urban environment.

"If people in the Rocky Mountain West never gardened at all, contenting themselves with the natural splendor around them, it would surprise no one."

—Carole Otteson,
The Native Plant Primer

When you visit a part of the planet that hasn't been modified and controlled by human whim, it's easy to be aware of natural diversity. If you go to an ocean beach, you'll see seagulls. Perhaps jellyfish and kelp will float around you in the waves; the sand beneath your feet may quiver with the blow-holes of buried clams and sand crabs. The sand may rise into dunes topped above the watermark by waving eelgrass and dune grass.

If you walk in a redwood forest instead, the air will be moist and cool under the canopy of giant evergreens, as moisture condensing from coastal fog falls in drops from the green needles. At your feet, ferns and huckleberry will stagger their stems in the half-light, and the western banana slug may crawl across your shoe, extending its pale yellow body a full nine inches and leaving a trail of sparkling slime.

Scientists call these delightful pockets of varied nature *biomes*. Biomes are regional areas defined by climate, topography, and soils, along with the specific plants and animals that have evolved within the limits of what the region can offer in terms of life support. Biomes cover broad stretches of territory that we identify as prairies, mountains, coasts, deserts, woodlands, and so on.

Environmentalists look at biomes close up, and call them ecosystems. The scientists who study ecosystems are fascinated by the animals, insects, and plants that live interdependently in a certain area.

Take the banana slug, for instance. Gardeners in redwood country generally hate the pesky things, so squishy underfoot, getting into the kitty bowls, plastering themselves on damp windows—and eating just about any kind of garden plant. Tomato seedlings, petunias, azaleas—slugs will devour them all overnight.

But there's one thing that banana slugs don't eat. Laboratory experiments have shown that banana slugs would rather chew on the cardboard box that holds them than munch on a redwood tree seedling. In its redwood forest biome, the banana slug has a role to play: It consumes the seedlings of competing tree species, and leaves the ground clear for baby redwoods to grow.

Gardeners are usually so happy creating artificial biomes (petunias + sprinkler + sea kelp fertilizer + lawn chair + poodle puppy, etc.) that they ignore the actual biome they happen to be living in. But while all gardens are the gardener's fancy, biomes are just plain fact. Your local weather is a fact; the earth beneath your paved roads and sewer lines is a fact. The bugs you swat today will show up again, because they belong in your ecosystem.

See for Yourself

1 What biome do you live in? What plants, birds, and insects are native to your neighborhood? Places to find this information include nearby natural history museums and zoos, local high school science teachers, and national organizations concerned with issues of habitat and ecology, such as the Sierra Club, the Audubon Society, and the National Wildlife Federation. Pictorial field guides for birds and insects can help you identify species common to your area. (See Sources.)

2 Visit an arboretum to discover which trees and shrubs can thrive in your locale. Arboretums are like tree museums—the specimens are all labeled. If you are shopping for a landscape tree, you can get a better idea of how a certain species will fit in your garden by viewing one that's fully grown. Mature plants, especially trees, often look quite different from skinny nursery specimens or mail-order catalog photos.

3 Stand in your garden and throw a ball over your shoulder. Where the ball lands, dig up a square foot of soil, plants and all, using a square-edged shovel to disturb as little of the earth as possible. Place the soil on a tarp on a table and dissect it, using pruners, tweezers, and a hand lens. Put any insects you find in a jar; take a sample of one leaf from each plant and tape or glue them to a piece of paper. How many different plants are there? How many different kinds of insects? Is the soil damp or dry? What is the weather like? Record your observations.

Even in the most urban environment, you can't escape the natural world. If you walk on 8th Street in Manhattan, you are part of the Manhattan Island ecosystem, a silty mound swirled around igneous lumps of a one-time volcanic plain, an island once clustered with stands of hemlock, oak, and maple, alive with natural springs. The trees may be gone, but the springs are still there. Spring-fed Minetta Creek is invisible, because it now runs in a subterranean passage beneath the buildings on 8th Street's southern side. When Jimi Hendrix recorded his memorable guitar solos of the 1960s at Electric Lady, a recording studio on the block, engineers claimed it was the resonance from the water channel beneath the studio that provided such a rich and full sound. Invisible or not, the natural environment still has an influence on city life.

Native Plant Gardens

Gardening with your biome in mind requires a certain mindshift on the part of the gardener: You learn to embrace the bugs, the slugs, and the plants you once considered weeds as interdependent partners in your landscape. You will not have tomatoes if you swat away all the big, black bumblebees coming to pollinate the flowers; you will not see any butterflies among your flowers if you spray or crush all the caterpillars you find munching on the leaves of your carrots and nasturtiums. If you wish to invite birds for their song and beauty, you must give up certain pesticides or risk poisoning the insect food that birds seek.

Bumblebee

A new style of gardening that is popular today is the native plant garden. Some folks are drawn to indigenous species once they realize how many pretty flowers come from "unimproved" wild plants. Some people like the idea of restoring a bit of their biome's natural ecosystem and feel in this way they can help conserve the diversity of life on this planet. Others find that planting a good roster of native trees and shrubs that bear nuts, berries, or nectar flowers is a sure-fire way to attract birds and butterflies to the garden.

You can get started with native plants by devoting a small section of your yard, or even part of a flower bed, to growing native perennial and annual flowers. Patience is required: Whether you start with wildflower seeds or starter wildflower plants (both available mail order, see Sources), it usually takes about three years for perennial wildflowers to establish themselves for a good stand of bloom. Some wildflowers are biennials, such as Texas bluebonnet, a form of wild lupine, which makes growth the first season, overwinters, and blooms the following spring. Annual grasses and wildflowers must be allowed to reseed: the seed stalks can look a little raggedy, but they provide food for birds. In early spring, to allow

---

The key to finding the right plants lies in knowing their origins. ...Plants native to regions of the North American continent where climactic extremes are commonplace have evolved in ways that minimize damage. After a particularly vicious hailstorm one late July, I stood forlorn among the icy white golf balls, green tatters and slush, immobilized and sickened by what fifteen horrible minutes can wreak on a garden. After a few stiff drinks and condolences from gardeners who had been spared, I was ready to walk around and survey the damage....

After three days I had two new compost piles—at least the shredded plants could feed next year's garden. I also came away with a remarkable discovery: the native plants in my garden had barely been scathed. Calliopsis *(Coreopsis tinctoria)*, mealy-cup sage *(Salvia farinacea)*, and prairie coneflower *(Ratibida columnifera)* among others all withstood the barrage from the sky. All originate from regions prone to hail. Their leaves were either too fine and narrow to be torn to smithereens, or they were pliable enough to bend with the impact, or in some cases, they were so strong and leathery that the hailstones slid or bounced right off them.

—Lauren Springer, *The Undaunted Garden*

---

perennials to spread more, the wildflower patch can be mowed or scythed to a few inches.

If your wildflower patch is part of a flower border, you can always cheat a bit and include some showy, hybridized versions of wild perennials, or sow some compatible annuals, such as poppy or bachelor's button, the first year. Most "canned" wildflower seed mixes contain such flowers; genuine wild annuals are usually more successful at self-seeding to continue the show into future years.

If you live in the United States, you can get a list of the wildflowers native to your state by contacting the National Wildflower Research Center in Austin, Texas, which was founded in 1982 by former First Lady, Ladybird Johnson. The center acts as a clearinghouse for information on native plants, offering nursery sources and a place to do research. The center also maintains a few acres replanted with Texas prairie grassland, so folks can drive by and see what the country looked like when buffalo, not beef cattle, roamed the Texas range.

If you plan to establish a native plant garden, you may need to educate your neighbors as well as yourself. In Wisconsin, prairie restoration is so much a part of the gardening aesthetic that front lawns in Milwaukee are just as likely to be planted with meadowsweet and shooting-star as they are with bluegrass turf. Even so, Wisconsin native Beatrice Smith, co-author of the book, *The Prairie Garden*, recommends checking with your neighbors before you start what some folks may view as merely an unkempt patch of weeds.

"At first, the neighbors may not understand," said Smith. "But people have discovered that once you get a prairie garden established, you don't have as much upkeep as with a regular flower garden. And you do get beautiful flowers."

Gardeners who have success with wildflowers often move on to native shrubs and native trees. Local species are the best adapted to the climate, and grow relatively trouble-free. In regions where summer droughts or heavy winter snowfall are common, native plants do not need summer irrigation or winter protection once they are firmly established, root to soil.

What Is a Weed?

When you stop to consider the thousands of plant species that have been introduced into this continent from faraway places over the last 500 years, it is amazing to realize how relatively few have escaped from cultivation to become invasive plant pests. There are perhaps two dozen worst offenders, ranging from the weedy, fire-prone eucalyptus trees that colonize California hillsides, to bittersweet and kudzu vines that suffocate recovering forests along the eastern seaboard.

Most plants are so specific to their native biome conditions that they require a great deal of fussing, feeding, and general encouragement—if not heroic measures—from the gardener to get a foothold, let alone thrive. No wonder some gardeners begin to resent the seemingly effortless growth of unwanted weeds.

Thoreau once termed a weed as a "plant whose virtue had yet to be discovered." To that I would add, we also call "weeds" those plants whose values have been forgotten. Most of our imported weeds were not originally considered pests. You may resent the dandelions and the wild blue chicory that are so difficult to dig out of your lawn. But European settlers brought those plants here deliberately, for their tasty spring greens and roots, which were ground as the home-grown substitute for coffee.

"I really get incensed when people call wildflowers and grasses weeds. My own philosophy is anything that volunteers and looks pretty gets to stay."

—Beth Blair,

Texas Master Gardener

Other plants we call weeds were unknown to European culture, but had a long history of use by Native Americans. A good example is the cattail rush, which will slowly crowd out competing species to completely overtake an untended garden pond. Native Americans harvested young shoots, pollen, and roots of the cattail for food; the rugged and waterproof stems were turned into mats, baskets, and cordage, even bundled into the stuff of temporary huts. In dry seasons, colonies of tall cattail stalks were easy to spot from a distance, a useful indicator of sites where groundwater lay invisibly and could be dug out to sustain a traveler's thirst.

Many of our native weeds have been hybridized into fancy flowers. Goldenrod is a popular garden flower in Germany and Britain, although it is despised at home, and shunned unfairly as an allergy plant (the real culprit is ragweed, a less showy flower that grows in similar conditions).

My favorite definition of a weed is "a plant in the wrong place." This allows for sufficient discretion on the part of the gardener. It is your yard, after all.

Attracting Wildlife

The ecology of your backyard includes its animal life: insects, birds, reptiles, mollusks, fish, and mammals—from the shy little shrew to your own pet dog. A garden that's planned to suit not just the flora but the fauna of your biome is called a habitat garden.

Like native plants that suddenly reappear in restored biomes, animal life slowly returns to reconstructed habitats. First insects come back—and you may not even notice the change. Insect predators follow, including small reptiles, birds, perhaps frogs. Sometimes fish re-appear in restored streams and swamplands; no one knows exactly how they arrive.

To attract animals to your garden biome, you can provide three things they need: food, water, and shelter.

"Some purists would have one believe that to plant any exotic is heresy. That's absurd. The act of creating a garden is a disturbance. ...A purist, to avoid hypocrisy, should remove the house along with the existing plants...."

—Lauren Springer, The Undaunted Garden

Providing food doesn't necessarily mean putting up a bird feeder. In fact, if you have plants that will provide berries, nuts, or seeds during the winter season, bird feeders are usually not needed to attract birds to your garden except in the coldest climates.

The birds that visit your backyard are serving an ecological function as well as adding to the color and beauty of your garden. Fruit-eating birds travel in flocks, and will often strip one entire tree of its berries before moving on to the next, or to a tree in another backyard. The flesh of the fruit is consumed, but the seed within is excreted once the bird has moved on; in this way the flocks help to distribute seeds over a broader area, increasing the odds for cross-pollination as opposed to self-pollination.

Seed-eating birds do a different job: they counterbalance some plants' efforts to reproduce far more seeds than could ever be accommodated in the ground around the mother plant. Squirrels that bury acorns carry these seed-cases far from the mother tree; if they forget where their treasure lies, the acorn has a chance to sprout into a new oak, usually at a good distance from its starting point.

Animals are also drawn to a water source. Migrating monarch butterflies will stop by your yard if you provide them with a mud wallow, a damp spot where they can draw moisture from the soil through the tubes in their long, curling tongues. A pond or birdbath becomes a magnet for many creatures.

Animals will stick around if you supply them with homes as well. Birdhouses and bat houses are one possibility. Old fallen logs are even better; until they crumble into humus, they provide a base for insects and shelter for smaller vertebrates. If you let a tall stump remain as a snag, a dead tree in your backyard

Woodpecker

can become an asset, serving as a potential target for woodpeckers or a perch or nesting site for raptors, such as owls and hawks.

When designing for wildlife, think in layers of greenery: tall tree canopies suit some bird species, while others seek the protection of dense evergreens, such as holly or brushy pines. Shrubs of different heights are inviting to birds as cover and lookout perches. Areas of the garden where it is difficult to grow vegetables or roses can be left undisturbed for creatures that nest in the ground.

Create a Butterfly Habitat

A butterfly garden can fit in any small garden space. You can even make it in containers for your patio. The first step in designing a butterfly garden is to find out which butterflies live in your area and what they eat. (For butterfly guidebooks, see Sources.) To attract butterfly visitors for just one summer, all you need to provide are nectar flowers. Garden flowers in the daisy family *(Compositae)* and carrot family *(Umbelliferae)* are good nectar plants; shrubs such as buddleia, often called butterfly bush, attract many kinds.

To create an inviting habitat that will encourage butterflies to stay, mate, and reproduce in your garden, you must plant not only nectar sources, but plants that will serve as food for hungry caterpillars along with plants that can serve as egg-laying sites.

A pretty yellow butterfly that is easily accommodated is the anise swallowtail, which will lay its eggs on wild fennel, Queen-Anne's-lace, parsley, and carrot foliage. In early summer, you can find its eggs, which look like pinhead-sized balls of creamy white. These hatch into caterpillars that are yellow-green with black dots and stripes; they grow quite large.

Try not to freak out when you see these monstrous worms munching on your carrot plants. Simply move the caterpillars to your designated habitat plants, such as a stand of wild fennel.

When they have eaten enough, the caterpillars turn into chrysalids, which hang like curled, dried leaves on the twigs of other garden plants. You can bring in the twigs with the chrysalids to hatch them indoors and watch the butterflies unfold, which takes about a week. Once they hatch, release the butterflies outdoors as soon as possible. Most will perch on your finger for a ride outside. Then wave to them as they fly away.

Good Bugs, Bad Bugs

You can practice habitat management on a small scale in your vegetable garden. Little kids often learn about good bugs and bad bugs: nearly every child recognizes the ladybug or ladybird beetle, with its lacquer-red wings dotted with black spots, and knows it will not bite or hurt and should never be killed.

When we are older the lesson must be expanded. Most adults still recognize the ladybug, and quite a few know the truth that is dear to the gardener's heart. Gardeners value these pretty little beetles because each one can devour armies of aphids, the pale green "bad bugs" that suck out the juices from our rosebushes.

Well, an aphid (or aphis) can be black, pink, or red as well as green, and there are about 80 different species of ladybug, some black, some yellow, and quite a few with no spots at all. If you want to have ladybugs around, it's also important to learn to recognize the immature or larval form of the ladybug, so you don't swat or spray it by accident. Baby ladybugs look sort of like quarter-inch-long alligators, with a roughened hide and stripes or spots. They eat aphids, too, as do the green lacewings (also known as aphid lions).

Adult ladybug

Juvenile ladybug

There are lots of other good bugs, and they're not always as pretty as the ladybug. The ground beetle is a pronged fearsome creature, shiny black as Darth Vader. It won't bite you, but it does attack smaller insects and eats the eggs of garden slugs. Some winged helpers are so small that you may not even notice them, such as tiny parasitic wasps and syrphid flies that prey on the larvae of whiteflies and fruit moths.

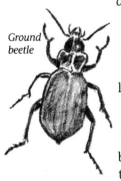

Ground beetle

Today you can order beneficial insects from a catalog (see Sources). A bag of fresh, live ladybugs costs about $5.95; they may be either farm-raised or gathered wild from some field in Colorado. Unfortunately, if the Colorado beetles don't like the plants in your backyard biome, they will fly off—if they don't starve to death.

The garden bug industry has answered this problem by developing hybrid ladybugs that stay where you drop them, and selling seed kits for plants that ladybugs like to live on, and yeast-based "ladybug food" that you spray on garden plants.

Government-run agricultural stations sometimes release beneficial insects to control bad infestations of crop pests. A recent example was an outbreak of whitefly throughout California. In selected areas, predatory beetles and a species of small wasp were released to control the whitefly, and home gardeners were requested to refrain from spraying whiteflies during the time it might take the beneficial insects to reach a critical mass and restore insect balance.

The ecological impact of moving bugs around should not be taken lightly. Some years ago, a mail-order nursery specializing in tropical plants offered packages of tropical butterfly chrysalids to their customers across the U.S. In the hue and cry that followed, it was decided that releasing butterflies outside of their natural range would, at the very least, doom the newly hatched butterflies to certain death away from their normal food plants and, at worst, create foreign competition to indigenous butterfly species. The product was withdrawn.

Organic gardeners claim that another way to attract beneficial insects to your garden is to plant certain types of flowers and plants. Much of this lore (and what empirical evidence exists) comes from an older tradition called "companion planting."

John Jeavons, one of the pioneers of modern organic gardening in America, recommends planting daisy-family flowers (*Compositae*) among food crops to attract beneficial insects. He also suggests that plants with pungent essential oils, such as garlic, nasturtium, carrot, and radish, may actually repel insect pests by their scent. One of the tenets of the biointensive gardening that he preaches is to mix crops in rows and beds, instead of devoting entire fields to a monoculture of a single food crop that may be more quickly overrun by a sin-

Nasturtium

gle pest or disease. In this case, the vegetable bed itself becomes a habitat garden of varied plants, and its insect life is resultingly more varied. For more about biointensive gardening, try Jeavons's excellent books. (See Sources.)

Pest Control by Nature's Calendar

Once you know the rhythms and growth cycles of your particular biome, you can use this information to protect your garden from insect invasions and many plant diseases. This is because as live organisms, insects and diseases also have their seasons. Here's one example where knowledge was power.

In the middle of July, Bob Chapman, the gardening columnist for the *San Jose Mercury News* in California, got a letter from a reader who had noticed scale insects dotting the leaves of her orchid plants. No amount of spraying industrial-strength malathion, she wrote, would dislodge them.

The sage and wise columnist gave her the right answer: all she could do at that point was to scrape the hard-shelled little beggars off the leaves with a nail file or wooden ice cream stick.

As Chapman noted, "Malathion is only effective on the immature stage of the insect. When the young crawl out from under their mother's shell and seek a new site they do not have the protective covering of the adult. After they have inserted their beak into the tissue and start feeding, they form the hard shell, which is impervious to most pesticides, including malathion."

Since these insects go through three or four generations in a single year, scraping off and removing mature insects will reduce future spread of the pest. The concerned reader might also monitor her orchid leaves in the future for tiny yellow crawlers (the immature form of the scale insects), and spray only when she sees them.

I'd also suggest that the concerned orchid-grower take a closer look with a hand microscope at the scales to see if they are alive or dead before she scrapes them off. Tiny pinprick holes in the shell of a dead scale would indicate the emergence of a scale parasite—an even tinier insect that's a natural control, usually available in

Alternative Disease and Pest Control

R esearchers at agricultural universities are pursuing alternative and non-invasive methods of pest and disease control. One of the most successful is the recent discovery that common baking soda can be used to make a homemade spray that will combat black spot and other fungus diseases of roses. The discovery was first made by commercial rose growers in Israel who were searching for an alternative to chemical rose sprays.

This research was replicated and refined in the U.S. at Cornell University. The resulting recipe has spread throughout the world's community of both amateur and professional rose growers and is sometimes called the Cornell Solution.

Mix together:
1 gallon water
1 tablespoon baking soda
1 teaspoon light horticultural oil (such as Eco-Oil)

Spray at weekly intervals. Because of its proven effectiveness, this is one of the few home remedies recommended by state and county agricultural agents and Master Gardener programs.

Another useful tool from the kitchen is aluminum foil. This can be used in place of the silvery reflective mulches now used at farms that grow strawberries, broccoli, and other crops susceptible to whitefly.

According to University of California researchers, the flying insects see a reflection of the sky in the shiny surface and become disoriented. They don't know which end of a plant is up, so they don't land there.

When you first notice whitefly on vegetables or annual flowers (they will fly up in a dusty cloud when leaves are disturbed), try a mulch of flat aluminum foil at ground level. Flattened aluminum foil or a pie plate will work. Remove the mulch in a week or two, or when temperatures rise in summer, to prevent overheating of the plants.

enough numbers to keep down the populations of scale in a garden. Unfortunately, such tiny beneficial insects are usually the first to die if sprayed with a chemical insecticide. By repeatedly spraying with malathion, our poor reader may have unwittingly killed off her best allies in the war against scale.

At Bunker Hill, the patriots were told not to shoot at the British "until you see the whites of their eyes"—in other words, until the enemy soldiers had marched close enough to be a good target for the amateur troops whose ammo was scarce. Gardeners, too, should hold off on spraying until they are sure the time is right.

There are times when the best you can do for a garden plant is to nuke it with the most powerful insecticide available. Applying insecticide at the proper time may mean you only have to do it once. But if you do other chores, also at the right season, you can get results with a less-toxic spray alternative—or you may not have to spray at all.

This approach to pest control is called *integrated pest management*, or IPM. IPM combines the best of what we know about managing life cycles of plants and their pests with the widest range of palliatives and cures—including everything from baking soda, which has recently been found to cure black spot on roses, to modern streptomycin compounds that specifically kill the bacteria that cause fire blight in pears.

IPM has been practiced for centuries without its fancy name. In medieval times, European peasant farmers burned the stubble in the fields once the wheat and barley were harvested. In this way, they killed off insect pests and disease spores clinging to the plant remains.

Most of the garden chores we consider seasonal hark back to traditions like that one. As a gardener, you honor your agricultural predecessors every time you rake away the fallen leaves underneath your apple tree or camellia bush. With the leaves, you are raking away the sleeping pupae of the codling moth and the microscopic spores of camellia petal blight, which are hiding in the debris.

The tradition we call "dormant season spraying"—an application of pesticides during the resting phase of a plant—dates back just a few hundred years, when mineral treatments to fruit trees included Bordeaux Mixture, an entirely "natural" but highly poisonous pesticide cocktail containing arsenic. Today's organic gardeners still use mineral compounds that contain sulfur and copper to inhibit bacteria and fungus diseases. These mineral compounds are available as dusts or liquids that can be mixed with horticultural oils that are usually petroleum-based. The oil itself offers a benefit to the fruit tree; it coats the limbs and smothers overwintering aphids and other fruit-tree pests.

Under nature's calendar, wind-borne spores that cause rust in apples—and brown rot, peach leaf curl, and shot hole fungus in stone fruits—happen to be traveling in the moist air of late winter, just at the same time the buds are swelling in your apple, peach, plum, and cherry trees. The fungi enter like an infection through the soft tissue of buds, blossoms, or young leaves. Splashing spring rains may move fungus spores from an already infected branch to new sites. (Shot hole fungus requires 24 hours of continuous wetness to penetrate plant tissue; apple scab can latch on in less than 10 hours on a warm, humid spring day. See page 137.)

Fungicides that contain the mineral copper or synthetic treatment products labeled for these problems kill the spores. You will be on target for your IPM if you spray the fungicide during the dormant period, and then again as soon as you see color in the flower buds. (Some folks call this the "popcorn" stage, because that's what the swelling whitish buds look like to them.)

Follow label directions to the letter (I do mean absolutely, please, and wear protective gloves, goggles, and a breathing mask) and spray directly onto the flower buds. If you had fungus problems the previous year, you will probably want to spray twice more, once when the flowers are fully open, and again at petal drop.

Apple blossom

Great Moments in Pomology

Back in 1954, a Cornell University researcher named W. D. Mills decided to track down exactly how long it took the spores of apple scab fungus *(Venturia inaequalis)* to penetrate and infect the soft growth on apple trees. Devising a chart that correlated temperatures with the length of time a leaf was wet, Mills found that, on average, spores became active when the temperature began to hover in the 40s. At that temperature, it would take about 30 hours for the spores to take hold. On balmy spring days, when the weather was between 50° and 75°, the spores could take hold in as little as nine hours (although hot weather over 75° slowed the process some 20 percent). This empirical research proved the how and why of what apple farmers have always known: if you need at least a sweater when you walk under the apple blossoms in springtime, you get more market apples come the fall. In cold weather, the apple scab fungus has less of a chance to establish itself.

The corollary to this lesson is that once leaves have grown out and fruits begin to form, no amount of spraying with anything will help keep the fungus from ruining the fruits. The only solution is to cut out infected fruits, leaves, and branches as you see them, and mark your calendar with a resolution to spray when the right time comes around again.

Whether you are worried about scales on your orchids or fungus in your plums, it's important to think about timing. Spraying the right thing at the wrong time doesn't do you any good—and may do your garden some harm by killing off your allies in the battle against bad bugs.

"Organic" Pesticides

J ust because a pesticide is labeled "natural" or "organic" doesn't mean it can't hurt you. Check the back label of any pesticide product and you will see that even the most highly touted organic cures carry cautionary statements that will raise the hair on your arms.

Mineral-based products with copper, sulfur, or hydrated lime can blind you; goggles are highly recommended when blending or spraying these. Rotenone and pyrethrum, two insecticides derived from plants, are potent nerve poisons that can be absorbed through the skin; don gloves and use care.

The modern organic arsenal also includes spray-on beneficials such as Bt *(Bacillus thuringiensis),* a minute bacterium that attacks all sorts of sod worms, caterpillars, and soft-bodied insect larvae. Though harmless to humans, Bt is indiscriminate—it will kill all the caterpillars that become butterflies, too. If your pest target is the green looping worm of the cabbage moth, spray only your cabbages, not the whole yard.

One of the selling points for modern synthetic pesticides is that some compounds are designed to break down into non-toxic molecules, often within minutes or hours of application. They can be used up to the day of harvest for food crops, and if used properly—on a limited plant area, and only for a specific disease or pest—the impact on the environment can be minimal.

In all cases, read the label before choosing or using any pesticide.

Dealing with Critters:
The Downside of Habitat Gardening

An inevitable result of bringing a garden into ecological balance is that you will get many unexpected visitors of the four-legged, six-legged, and eight-legged variety. It's not so difficult for humans to get over a fear of spiders, or stop leaping for the pesticide spray at the sight of a strange bug. But when the deer eat up the rosebushes, and the cottontails consume the fine French lettuce, even the most animal-loving gardener may be forgiven for entertaining thoughts of shotguns and traps.

I mentioned earlier that animals will be attracted to a food source. Well, if you provide a tasty buffet of natural and imported greenery, animals will come and get it. This is not just a rural problem. When fields and wildlands are developed for suburban housing, displaced critters don't move or die off as much as they adapt and find new places to hide. You may wake up in the morning to find raccoons in your garbage cans, opossums in your bean patch, and deer tracks where you planted salvias the day before.

As hunting and trapping have become less popular, gardeners are seeing field and woodland animals more frequently. Their rising numbers also attract and sustain growing populations of native predators. After decades of near-extinction, mountain lions once again roam within 50 miles of downtown San Francisco and are already responsible for one human death. Eight-foot alligators emerge from Florida canals to carry off pet poodles, and more than one countryside commuter in the Northeast has been startled out of his breakfast coffee by the sight of a black bear in his backyard.

What can you do to keep nature in its place? Most gardeners are willing—some even eager—to share their landscape with fleet-footed herbivores, as long as the pretty deer don't munch all the special plants and flowers. You can have a wild garden, and a vegetable and flower patch too, if you isolate a portion of your garden for special crops. This is one alternative to surrounding your whole property with a 12-foot fence—an expensive proposition, and one that's perhaps at odds with the idea of "natural gardening."

In deer country, it is not uncommon to see small orchards of dwarf fruit trees completely surrounded by fencing. In my region, I know of at least one gardener who solved the wildlife problem by constructing a cage, 15 feet wide, 15 feet tall, and 30 feet long, with wire mesh fencing on the top and sides, to hold raised beds for vegetables and prized tea roses. With a comfy bench or two and a pitched roof, this might have made a pretty summerhouse.

Painting your wire fencing black makes it disappear when viewed at a distance. In the Berkeley hills, natural gardener Jana Drobinski grows her vegetables in a space ringed with sculptural curlicues of rebar, draped with near-invisible drifts of black bird netting. It is enough to keep the deer away, and Jana can still watch the animals; from her back deck the vegetable garden looks like a little sculpture area in the middle distance.

Other gardeners use hedging to keep animals out of select garden areas. One man I know mixes his rosebushes and cottage flowers in a border with tall, ripply ornamental grasses, and the backs of his borders are lined with even taller sorts, such as Miscanthus and Elymus, to discourage deer. The sharp edges of the grass blades "hurt their noses," he says; I suspect sudden wind-ruffled movements of the grasses and their waving seed panicles also spook the deer.

Yew (poisonous to deer) and privet (which bounces back from brows-

ing) are good choices for hedging, as are spiny plants like barberry and trifoliate orange. These make an impenetrable, thickly thorned barrier that keeps people out as well. The orange has the advantage of sweet-scented flowers and decorative orange fruits.

Barrier methods also work on a small scale. Birds can be kept out of newly seeded vegetable gardens by poking brushy twigs into the ground around the seeds. The theory is that birds may fear getting their wings caught should they need to take flight in a speedy getaway, and so avoid the brushy tangles.

The garden slugs and garden snails, which decimate marigold and basil seedlings, are effectively controlled by a barrier made of thin copper stripping. This trick has been used for about 25 years to protect orange groves in southern California from depredations by the European brown snail, an introduced pest known for climbing up citrus trees to eat leaves and make holes in citrus fruit. It is certainly an improvement on the previous method used to control snails in commercial groves, which involved placing strychnine-laced bait under the fruit trees. Copper's repellent effect was discovered when citrus tree trunks painted with liquid copper sulfate for an unrelated disease problem were found to be unusually snail-free.

"One for the cutworm, one for the crow, one for the cabbage worm, and one to grow."

Laura Ingalls Wilder, Little House on the Prairie

How does it work? According to Dr. Carl Koehler of the University of California at Riverside, there's something in slug and snail slime that has an electrochemical reaction when it touches copper, "like a small electric shock." Gastropods will not cross copper, and a thin wash of the metal will do. Today if you visit commercial orange groves you may notice each tree sports a copper-strip collar around its trunk.

Copper stripping is available in small packages from garden suppliers and remains effective until it oxidizes and dulls (about 12–18 months). It can be cut with scissors and is flexible enough to be wrapped around pots or stapled to the sides of wooden planters or raised beds. You can also make a circlet of copper and push it into the ground around plants especially attractive to slugs, such as hosta and delphinium. In my own garden, I surround lettuces and tender greens with a little copper corral. On spring mornings, I often find 30 or 40 snails with dazed expressions massed on the far side of the copper.

If you provide wild food for wild animals, and make it difficult to reach desirable plants, smarter wild animals can perhaps be "trained" to avoid certain areas. Managing wildlife in a garden environment basically means directing them elsewhere. Bird netting keeps animal thieves away from your fruiting cherries, as does hanging glittery wind chimes or some other kind of "scarecrow." A new method to keep squirrels from raiding bird feeders is to mix birdseed with finely ground chili peppers; apparently birds can't taste the chili, but squirrels can—and they get a mouthful they don't forget quickly.

Gophers who tunnel under lawns and flower beds in search of earthworms in your healthy soil can be foiled if you install wire mesh under the planting areas. Wire gopher baskets in various sizes are used by professional landscapers when transplanting trees and edible bulbs, such as tulips. I recently saw a 1,000- by 1,000-foot lawn where the sod had been laid on a chicken wire grid six inches below the surface, to prevent gophers from marring the pristine look of the turf.

Consider this advice from a pest control expert: "To get rid of gophers, have lots of garden parties—they hate noise and move out if there is too much disturbance up above." Toothed and dangerous animals are best taken care of by your city's department of Pest Control (sometimes called Vector Control). Another highly effective deterrent to mammal marauders is a large, loud dog.

The Circle of Life

Critters can sometimes get in the way of gardening, but it helps to realize their families lived here long before ours came along. I had no access to my compost pile for two long months one spring-time, because ground-nesting bumblebees had made themselves a hive in the soft duff, and would rise up in menacing troops whenever I extended a hoe or trowel in their direction.

Nuking them with pesticide seemed a bit at odds with the ecological aesthetic of making compost in the first place, so I asked a gardening neighbor what to do.

"Let them be," she said. "They'll leave the nest in their time." By June, the bees had indeed departed, though I saw lots of bumbles that summer pollinating my tomato crop.

In a small garden you make choices. I don't chop down the stalks of a weedy wild anise (an imported European escapee) because its greenery provides food for the caterpillars of the large yellow anise swallowtail butterfly. However, I do clip off the seed heads before they ripen and blow across everybody's yards. There is plenty of anise here already, and it's a pain to grub it out once it reaches a six-inch height.

Mockingbird

For three years in a row, a family of mockingbirds has nested on my roof, and the birds do a good job of eating garden insects, although I'm always saddened when I see they've gotten one of the big green swallowtail caterpillars.

Hummingbirds visit daily, attracted by beds filled with native salvias, penstemons, and a pink-flowered wild currant. They crisscross a sky also trafficked by several different species of dragonfly who seem to enjoy the birdbath. (My favorite species has a body checkered yellow and black, just like a New York City taxicab).

I've gotten used to the mice—a little scurrying movement underneath a row of pots at the scrape of a moved patio chair, a rustling sound heard just above the noise of the freeway. Each winter I wonder if that hawk will be back.

Sources: A Resource Guide for Readers

Chapter 1: Plants and People

Alein, Vicki Herzfeld, and Lee Reich. *Uncommon Fruits Worthy of Attention: A Gardener's Guide.* Menlo Park, CA: Addison-Wesley, 1991.

Ashworth, Suzanne. *Seed to Seed.* Decorah, IA: The Seed Savers Publications, 1995.

Hansen, Richard, and Frederick Stahl. *Perennials and Their Garden Habitats,* 4th edition. Translated from the German by Richard Ward. Portland, OR: Timber Press, 1993.

Jefferson, Thomas. *The Garden and Farm Book of Thomas Jefferson.* Edited by Robert C. Barron. Golden, CO: Fulcrum Publishing, 1987.

Neal, Bill. *Gardener's Latin: A Lexicon.* Introduction by Barbara Damrosch. Chapel Hill, NC: Algonquin Books, 1992.

Reddell, Rayford. *Growing Good Roses.* Petaluma, CA: Garden Valley Press, 1988.

Staff of the L. H. Bailey Hortorium, Cornell University. *Hortus III.* New York: MacMillan, 1976.

Thomas, Graham Stuart. *The Graham Stuart Thomas Rose Book.* Portland, OR: Timber Press, 1994.

_____. *Ornamental Shrubs, Climbers and Bamboos: Excluding Roses and Rhododendrons.* Portland, OR: Sagapress/Timber Press, 1990.

_____. *Perennial Garden Plants: Or the Modern Florilegium: A Concise Account of Herbaceous Plants, Including Bulbs, for General Garden Use.* Portland, OR: Timber Press, 1990.

Virgil. *The Georgics.* Translated by L. P. Wilkinson. New York: Viking Penguin Classics, 1983.

White, Katherine. *Onward and Upward in the Garden.* New York: North Point Press, 1997. (Essays originally published in *The New Yorker.*)

Whittle, Tyler. *The Plant Hunters.* Philadelphia, PA: Chilton Books, 1970.

Russian fruits, pawpaw, pluots, and apriums are available through Raintree Nursery, 391 Butts Road, Moreton, WA 98356. (360) 496-6400.

For information about unusual and native fruits, contact North American Fruit Explorers (NAFEX) c/o Vorbek, Route 1, Box 94, Chapin, IL 62628.

Heirloom seeds are available through Seeds of Change, 621 Old Santa Fe Trail, #10, Santa Fe, NM 87501. (505) 438-8080.

The Seed Savers Exchange, 3086 North Winn Drive, Decorah, IA, 52101. (319) 382-5990.

Chapter 2: Inside Plumbing

"Determining Daily References for Evapotranspiration," U.C. Agricultural Pamphlet #21426. Oakland, CA: ANR Publications, University of California. (Available by mail from ANR Publications, 6701 San Pablo Avenue, Oakland, CA 94608.)

Fishman, Ram. *The Handbook for Fruit Explorers.* Chapin, IL: North American Fruit Explorers, 1986.

Hill, Lewis. *Fruits and Berries for the Home Garden.* New York: HarperCollins, 1992.

Kourik, Robert, and Heidi Schmidt. *Drip Irrigation for Every Landscape and All Climates: Helping Your Garden Flourish While Conserving Water!* Santa Rosa, CA: Metamorphic Press, 1993.

Lloyd, Christopher. *The Adventurous Gardener.* New York: Random House, 1983.

Thompson, Peter. *Creative Propagation: A Grower's Guide.* Portland, OR: Timber Press, 1992.

Yang, Linda. *The City and Town Gardener: A Handbook for Planting Small Spaces and Containers.* New York: Random House, 1995. (Originally published as *The City Gardener's Handbook*).

Agricultural information and publications office, University of California at Davis, University Services Building, Rm 110, Davis, CA 95616. (916) 757-8930.

Grafting supplies available through Raintree Nursery catalog. (See listing under Chapter 1.)

Chapter 3: Know Thy Dirt

Brown, Martha. "Dig Deep to Make Your Garden Bloom," *Exploring,* vol. 20, no. 1 (Spring 1996). (*Exploring* is a quarterly magazine published by the Exploratorium, 3601 Lyon St., San Francisco, CA 94123.)

Jekyll, Gertrude. *Wood and Garden.* Suffolk (U.K.): Antique Collector's Club, 1981. (First published in 1896.)

Lawrence, Elizabeth. *A Southern Garden.* Chapel Hill, NC: University of North Carolina Press, 1991. (First published in 1942.)

Miller, Mary K. "Turning Garbage Into Gold," *Exploring,* vol. 20, no. 1 (Spring 1996).

Mitchell, Henry. *One Man's Garden.* New York: Houghton Mifflin, 1992.

For soil testing kits and services, contact Peaceful Valley Farm Supply, PO Box 2209, Grass Valley, CA 94945. (916) 272-4769.

Fava bean seeds are available from Thompson & Morgan, PO Box 1308, Jackson, NJ 08527-0308.

Chapter 4: Cycles and Seasons

Gallston, Arthur W. *Life Processes of Plants.* New York: W. H. Freeman and Co., Scientific American Library, 1994.

Hale, Judson D. *The Old Farmer's Almanac.* New York: Random House, published annually.

Jeavons, John. *How to Grow More Vegetables: Fruits, Nuts, Berries, Grains, and Other Crops,* 5th ed. Berkeley, CA: Ten Speed Press, 1995.

von Armin, Elizabeth. *Elizabeth and Her German Garden.* London: Virago, 1985. (First published in 1901.)

Yamane, Linda, and Glenn Keator. *In Full View: Three Ways of Seeing California Plants.* Berkeley, CA: Heyday Books, 1995.

The Great Tomato Race database is maintained by the author, in a computer file and a cardboard box someplace.

Dahlia imperialis (giant tree dahlia) is not commercially available in the U.S. Starter plants are sold occasionally by Friends of University of California Botanical Garden, Berkeley, and by the San Mateo Garden Center, San Mateo, CA.

Black locust *(Robinia pseudoacacia)* is available through Forest Farm, 990 Tetherow Road, Williams, OR 97544. (207) 829-5830.

Chapter 5: Fooling Mother Nature

Lawrence, Gertrude. *Through the Garden Gate.* Edited by Bill Neal. Chapel Hill, NC: University of North Carolina Press, 1990.

Rooting hormone products are available at most garden centers.

Corkscrew willow is available through Forest Farm, 990 Tetherow Road, Williams, OR 97544. (207) 829-5830.

Chapter 6: Ecology and Habitats

Flint, Mary. *Pests of the Garden and Small Farm.* U.C. Publication #3332. Oakland, CA: ANR Publications, University of California.

Gordon, David G. *Field Guide to the Slug.* Seattle, WA: Sasquatch Books, 1994.

Holmes, Roger, ed. *Taylor's Guide to Natural Gardening.* New York: Houghton Mifflin, 1993.

Ottesen, Carole. *The Native Plant Primer.* New York: Harmony Books, 1995.

Smith, J. Robert, and Beatrice S. Smith. *The Prairie Garden: 70 Native Plants You Can Grow in Town or Country.* Madison, WI: University of Wisconsin Press, 1980.

Springer, Lauren. *The Undaunted Garden: Planting for Weather-Resiliant Beauty.* Golden, CO: Fulcrum Publishing, 1994.

Thompson, Bob. "Mailbag," *San Jose Mercury News,* Aug. 1, 1996.

Wilder, Laura Ingalls. *Little House on the Prairie.* New York: HarperCollins Juvenile Books, 1975.

National Wildlife Federation, 1400 16th St. Northwest, Washington, D.C., 20036. Write for information.

National Wildflower Research Center, 4801 La Crosse Ave., Austin, TX 78739. Write for information on native plant, wildflower, and regional seed sources.

Xerxes Society, 10 Southwest Ash Street, Portland, OR 97204. Offers information on butterfly gardening.

Beneficial insects and organic gardening products are available through Gardens Alive!, 5100 Schenley Place, Lawrenceburg, IN 47025.

Snail-Barr Copper Strips are available through The Natural Gardening Company, 217 San Anselmo Avenue, San Anselmo, CA 94960.

Index

dividing, 104
invasive, 33
water and, 32–35
Rootstock, defined, 46
Rotation of crops, 51
Rust, 15, 26, 27

Salt burn (leaf-tip burn), 50
Scion, defined, 46
Sealants, for wound care, 43
Seasonal affective disorder
(SAD), 87
Seasons.
See Growing seasons
Seeding, self-, ecology and,
121
Seeds
abscisic acid in, 110
energy stored in, 81
germination of, 78, 89
hybrid vs. open-
pollinated, 22–23
moonlight and, 94
starting, 111–113
Seed Savers Exchange,
118, 119
Shadows, winter angles of,
90, 91
Shield budding, 47
Shrubs.
See Trees and shrubs
Slugs, 131, 141, 142
banana, 121, 122
Snails, 141, 142
Soil, 52, 75
additives for, 67–72
composition of, 53–54, 58
double digging, 57
heavy metals in, 66, 67,
72
ion exchange between
plants and, 54–56
living organisms in, 58,
60, 61
optimum texture for, 56
pH of, 61, 64–67
testing, 62–63, 65
toxins in, 66–67, 72
types of, 56, 61
Solanaceae, 16–17
"Solvent drag," 55

Sow bugs (pill bugs), 60
Species, 12–14
Stamen and pistils, 17, 79
Stolon, 109
Stomata, 25, 26
Subsoils, 53
Sulfur, 65, 70
Summer fruit drop, 102, 103
Summer pruning, 42
Sunburn, 27, 49, 50
Sunlight, 79, 81, 90, 91
Superphosphates, 68
Sweet soils, 61

Taproot, 32–33, 37
T-budding, 47
Tetraploid hybrids, 21
Tip pruning, 100
"Topping," 41
Topsoil, 53
Totipotency, 103–105
Toxins in soil, 66–67, 72
Trace minerals, 70
Transpiration, 24–26
evapotranspiration (ET)
rates, 29, 30
transplant stress and, 34,
35
weather changes and, 28
Trees and shrubs
cambium of, 36
native, 126
pruning, 39–43, 46
vascular network, 37
Tropical plants, 8, 9, 88
Tubers, energy stored in, 81

Umbelliferae, 15
USDA climate zones,
114, 115

Varieties, within species, 12,
118
Vegetables
companion planting of
132–133
deadheading, 108
disbudding, 102
families of, 15
water requirements and
flavor of, 33

wilt diseases in, 51
Verticillum wilt, 51
Vitamin supplements, 70

Water, 24, 48
conserving, 47, 48
movement of, 24, 26
root systems and, 32, 35
soil amendments and, 47
sugars transported with,
80–81
transport system, 36, 37,
38
Watering
excessive, 32
fertilizing and, 55
indicator plants, 29
requirements for, 28–32
timing of, 26, 27
Watersprouts, 42
Weather
forecasting plants, 11
native plants and, 125
outwitting, 116–117
USDA climate zones and,
114–115
water needs and, 28
Weeds, 127–128
Weed whacking, tree dam-
age from, 36
Whip graft for apple trees,
44–45
Whiteflies, 132, 134
Wildflowers, growing,
125–126
Wildlife, attracting, 128–130
Wildlife pests, 139–142, 143
Wilting, recovery from, 28
Wind-indicator plants, 11
Winter
anti-transpirant sprays
in, 35
sun angles in, 90–91
See also Dormancy

Xylem and phloem cells, 36,
37, 38

Zinc, 70

Credits and Acknowledgments

Edited at the Exploratorium by Pat Murphy
Design by Gary Crounse
Production Editing by Ellyn Hament
Production by Stacey Luce

Image Credits
Photos by Amy Snyder and illustrations by Esther Kutnick, unless otherwise indicated.
Page 25: photo by R. F. Evert/thanks to Susan Eichhorn; page 26: photo by M. M. McCauley/thanks to Susan Eichhorn; pages 33 and 84: photos by Susan Schwartzenberg; page 94: photo by Katya Klasen/thanks to Laurene Vincen; page 104: photo by Ellen Klages.

Acknowledgments from Mia Amato
This book could not have been written without input from the hundreds of adventurous gardeners I have met as a newspaper columnist in San Francisco. Over the past five years, they have generously shared the benefit of their experiences. Thanks especially to Kurt Hase, Wally Hennessy, Andy Liu, Dennis Makishima, Chase Rosade, Christine Smith, Lorraine Vinson, Linda Yang, my mentors Dr. Bob Raabe and Richard Molinar of the University of California, and Charlie Mazza of Cornell University.

Thanks also to Hazel White, for her kind words; to Jo Mancuso and Frances La Rose, for their professional and personal support; and to Pat Murphy, who is a priceless pearl among editors, and who, along with Kurt Feichtmeir and the enthusiastic Exploratorium staff, made the daunting task of a first book one of the most enjoyable experiences of my writing life.

Acknowledgments from The Exploratorium
Special thanks to David Sobel for his editing expertise. Also, thanks to the many members and friends of the Exploratorium staff who contributed time, ideas, and expertise to this book. We would particularly like to thank Gary Crounse for his design expertise and Buddhist calm in times of great stress; Esther Kutnick for working heroically to complete the illustrations; Larry Antila for his courier services; Stacey Luce for her daring design rescue at the eleventh hour; Amy Snyder for her photographs; Mark Nichol for his photo research; Mary Miller for her suggestions along the way; Megan Bury for helping out with many tiny changes; Karen Kalumuck for reviewing the manuscript from the biologist's perspective; Kurt Feichtmeir for his budgeting acumen; Ruth Brown for her sage advice; Ellen Klages for her careful copyediting, and Ellyn Hament for her valiant attempts to catch every last typo.

About the Author

Mia Amato is a full-time journalist and a lifelong gardener who has successfully grown orchids in Manhattan and sweet corn on the foggy hills of San Francisco. She is the author of two syndicated newspaper columns, "Yardscapes" and "Mia's Kitchen Garden," and of gardening stories for magazines such as *House Beautiful, Fine Gardening, Garden Design,* and *Sunset.*

About the Exploratorium

The Exploratorium, San Francisco's museum of science, art, and human perception, is a place where people of all ages make discoveries about the world around them. The museum has over 600 exhibits, and all of them run on curiosity. You don't just look at these exhibits—you experiment with them. At the Exploratorium's exhibits, you can play with a captive tornado, generate an electric current, see what's inside a cow's eye, and investigate hundreds of fascinating natural phenomena.

Each year, over half a million people visit the museum. Through programs for teachers, the Exploratorium also encourages students to learn by asking their own questions and experimenting to find the answers. Through publications like this one, the Exploratorium brings the excitement of learning by doing to people everywhere.

Visit the Exploratorium's home page on the World Wide Web at:

www.exploratorium.edu

Next time you are in San Francisco, come visit the Exploratorium!